Studies in the Foundations
Methodology and Philosophy of Science
Volume 2

Editor: Mario Bunge

Coeditors: Peter G. Bergmann · Siegfried Flügge
Henry Margenau · Sir Peter Medawar · Sir Karl R. Popper
Patrick Suppes · Clifford A. Truesdell

Quantum Theory and Reality

Edited by

Mario Bunge

Springer-Verlag Berlin Heidelberg New York 1967

Mario Bunge

McGill University
Montreal/Canada

ISBN-13: 978-3-642-88028-5   e-ISBN-13: 978-3-642-88026-1
DOI: 10.1007/978-3-642-88026-1

Title-No. 6761

Contents

Introduction

MARIO BUNGE
The Turn of the Tide 1

The Epistemological Issue

1 Sir KARL R. POPPER
Quantum Mechanics without "The Observer" 7

2 HENRY MEHLBERG
The Problem of Physical Reality in Contemporary Science 45

ψ, Measurement, and the New Particle Crop

3 PETER G. BERGMANN
The Quantum State Vector and Physical Reality 66

4 HENRY MARGENAU and LEON COHEN
Probabilities in Quantum Mechanics 71

5 JEAN-PIERRE VIGIER
Hidden Parameters Associated with Possible Internal Motions of Elementary Particles 90

Objectivistic Foundations of Q.M.

6 GÜNTHER LUDWIG
An Axiomatic Foundation of Quantum Mechanics on a Nonsubjective Basis 98

7 MARIO BUNGE
A Ghost-Free Axiomatization of Quantum Mechanics 105

Introduction

The Turn of the Tide

During centuries physicists were supposed to be studying the physical world. Since the turn of the century this assumption has often been challenged as naive: it was proclaimed that physics is not about the external world but about observers and their manipulations: that it is meaningless to talk of anything else than observation devices and operations: that the laws of physics concern our knowledge rather than the external world.

This view of the nature of physical science has old roots in philosophy but it was independently reinvented by a number of philosophically inclined physicists, notably ERNST MACH. These scientists were disgusted with the school philosophies and they were alarmed by the increasing number of physical concepts which they regarded as metaphysical or beyond experimental control, such as those of absolute motion, ether, electromagnetic field, and molecule. Reasonably enough, they wished to keep physics testable. To accomplish this goal they adopted the safe method, namely to banish every idea that could not be closely tied to observation. In this way they certainly avoided the risks of untestable speculation but they also failed to enjoy the benefits of theoretical invention. Furthermore they instituted unawares a new metaphysics that was to dominate the philosophy of physics for half a century: the metaphysics according to which the world is made of sense experience.

The new philosophy — new to physicists, that is, not to philosophers — took on quickly. By 1900 all deep theories — all theories involving unobservables — had become suspect and even disreputable. In particular, atomic theories were neglected and even derided, for they failed to comply with the philosophical requirement that only observables should be handled. The work of BOLTZMANN — a resolute atomist and realist — was practically unknown while the word of MACH — the antiatomist and phenomenalist — spread like the gospel. Thermodynamics — or rather thermostatics — was regarded as the paradigm to be imitated. Physical theory was required to stick to observation, to relate empirical data and predict possible outcomes of experiments. The hypothesizing of hidden entities and inner structures became taboo. On the other hand sensible qualities — such as observability — became again as respectable as they had been before GALILEI expelled them from physics.

The observer-centered epistemology and the associated phenomenological or externalist approach to theory construction were in the air when atomic physics and relativity changed the face of our science. So much so that the few outspoken realists extant, such as BOLTZMANN and PLANCK, felt compelled to write in a defensive mood. So much so that the new discoveries made no impact whatsoever on the prevailing philosophy, even though atoms were not directly observable and relativistic covariance satisfied the realistic requirement of an observer-free theoretical science. Far from giving birth to a new philosophy, the new science adopted the language, if not the spirit, of the prevailing philosophy. The relativist got into the habit of saying that the rest mass of a point particle was the value an observer riding it would measure — or even the value the particle itself would see. And the atomic physicist learned to state that every atomic state was the outcome of some laboratory manipulation, the very idea of an autonomous external world being a metaphysical legacy from classical physics. Observers and observables — or rather the corresponding words — began to invade the whole new physics. Nobody seemed to notice the inconsistency.

By the mid-1920's the conversion of physicists to observationalism — or operationalism — was almost complete. The ideal of an objective (subject-invariant) physical theory was pronounced dead. Throughout, manipulations took the place of physical objects. Thus in his influential treatise on relativity, EDDINGTON [1] explained that symbols acquire a physical meaning insofar as they are correlated to measurement operations. He even claimed that "Physical quantities defined by operations of measurement are independent of theory, and form the proper starting-point for any new theoretical development". CARATHÉODORY [2] — who had shown operationalist leanings already in his famous paper of 1909 — when wondering what the physical meaning of certain thermodynamical functions could be, had in mind the way their numerical values could be measured. And HEISENBERG [3], in an epoch-making paper on quantum mechanics, declared that, contrary to classical physics, the aim of quantum mechanics was to relate only strictly observable quantities.

When BRIDGMAN'S book [4] came out in 1927, it became quickly popular because it said clearly and in detail what people had been thinking all along. (HUGO DINGLER had been saying pretty much the same thing for over a decade but physicists did not take him seriously because he opposed the new physics.) By that time the Verein Ernst Mach, or Vienna Circle, was founded as a sort of institutionalization of the *Zeitgeist* prevailing in scientific circles. Its members endeavored to evolve the philosophy of science that had been sketched half a century earlier by MACH, and they did it with unusual ability and intellectual honesty, by joining for the first time the empiricist tradition to modern logic: logical empiricism was born. From then on, for at least two decades, the philosophy of science became almost identical with logical empiricism. Among the few notable exceptions were ALBERT EINSTEIN — a former follower of MACH —, MAX PLANCK, ERWIN SCHRÖDINGER, and the philosophers EMILE MEYERSON, ERNST CASSIRER, and KARL POPPER — at that time the lonely philosopher who criticized positivists with their own logical tools and without school strictures. Most of the other antipositivists were inimical to logic and/or science.

For the first time since the golden 17th century, a group of brilliant philosophers agreed with most of the eminent physicists of their time about the nature and aim of physical science. The two parties agreed that what had been wrong with classical physics was that, having espoused a realistic epistemology, it focused on the object rather than on the subject's operations. They agreed that, when properly understood, the new physics reduced the physical object to little more than the grin of the Cheshire cat. What remained were observers and observables, and the latter were not real properties of autonomously existing things but mere possibilities of observation. Not observations on something out there but just observations. The external world was gone: only "its" representation was left. True, the operationalist would grant reality to his desk and eventually also to the atoms that make it up, but he means by 'reality' a set of human operations and apperceptions not the aggregate of things outside the human mind. Indeed, to the operationalist 'x is real' means "x is observable or measurable". It is not just that observability is regarded as a criterion of reality, as a test of the hypothesis that something exists: the operationalist identifies reality with operational possibility or even (radical branch) with operational actuality, thereby denying independent reality to his desk or at least to its atomic constituents. The empiricist variety of subjectivism should not be mistaken for the traditional philosophical idealism although it comes close to it: while the idealist — like JEANS and EDDINGTON in their old age — claims that things are merely ideas and, in particular, that only mind matters as regards matter, the empiricist denies both ideas and matter an independent existence.

This covenant between physicists and philosophers, achieved for the first time since the Middle Ages, lasted about two decades. During this period The Observer displaced both matter and God. But philosophers, if fond of logical analysis, cannot resist the temptation of critically examining philosophical assumptions, not even their own pet hypotheses. So, while physicists kept the crude operationalist philosophy of the 1920's, most of the philosophers responsible for it have ever since

1*

changed their minds about lots of things, mainly as a result of vigorous discussions and honest soul-searching. Today there are hardly any orthodox operationalists and phenomenalists left in the philosophical profession: operational definitions are acknowledged as deficient, and phenomenalism as inconsistent with the use of theoretical terms [5]. While few philosophers will call themselves realists and even fewer materialists, most of them have dropped subjectivism. The final reconversion to realism is practically a matter of form. As far as most philosophers are concerned, the world is more or less tacitly allowed to run by itself.

But contemporary physicists are die-hard positivists of the 20's and 30's: they have imbibed this subject-centered philosophy as undergraduates — and mind, not in philosophy courses but in physics courses. Almost every physics textbook decrees that no symbol is physically meaningful unless it is "defined" by a set of laboratory operations, and gives the impression that there would be no events without observers. While in the early 1600's the conservative philosopher would refuse to look through the scientist's telescope, as late as in the 1960's most physicists still refuse to use the logoscope built in recent years by the philosopher. Thus they will insist on interpreting every symbol in terms of human operations even though a logical analysis of the given symbol fails to disclose its dependence on the observer, and even if the operation itself is impossible — as, e. g., the measurement of a field strength at a given time and at every point on a given surface. For the first time in history, scientists have managed to outdogmatize philosophers.

Yet in recent years the tide has begun to turn, not only in philosophy but also in physics. Within the so-called Copenhagen school itself, certain objectivist trends are appearing. Thus while BOHR [6] had always insisted that the epistemological lesson of atomic physics was that the object could not be separated from the subject, HEISENBERG [7] has recently admitted that "it does not matter whether the observer is an apparatus or a human being", and that "the introduction of the observer must not be misunderstood to imply that some kind of subjective features are to be brought into the description of nature". The days in which it was believed that "ordinary (i. e., macroscopic) phenomena are in a way engendered by repeated observations" [8] seem to be over — well, almost over. The physicist of the latest generation is operationalist all right, but usually he does not know, and refuses to believe, that the original Copenhagen doctrine — which he thinks he supports — was squarely subjectivist, i. e. nonphysical.

Many physicists are beginning to wonder not only whether the current theories are sufficient, but also whether it would not be worthwhile to analyze and reinterpret them. Some heretics find obscurities and inconsistencies in them. Others go as far as wondering whether physical theories may not, after all, be about chunks of reality rather than about human actions: they begin to suspect that the concern or *referent* of a theory need not coincide with the way it is put to the *test*. Finally, others dissociate themselves frankly from the subjectivistic epistemology underlying the Copenhagen interpretation of the quantum theory. Thus WEISSKOPF [9] rejects HEISENBERG's claim that quantum mechanics "represents no longer the behavior of the elementary particles but rather our knowledge of this behavior", and maintains that the success of that theory "gives us confidence in having discovered something about the real world". Surely the word 'reality' is still shunned and various euphemisms are resorted to. Still, more and more people are wondering whether reality is not the skeleton in the cupboard of 20th century physics, and whether eventually the crime will have to be cleared up. They are beginning to suspect that the official philosophy of physics — which is no longer held by philosophers — had gone too far in its eagerness to dispel metaphysical inscrutables. Surely physical hypotheses must be susceptible to experimental test, but why should they not concern the external world and why should they not explain how things work?

After several decades, in sum, several people are starting to ask what became of reality and whether a new philosophy of physics should not be explored, one giving more credit to mother Nature and more freedom to the creative imagination of the theoretician. The new philosophy, it is felt, will have to avoid two extremes that failed long ago: dogmatic subjectivism and dogmatic realism. It will have to be critical and open-minded. It won't do to repeat that physics is about *physis*: the new philosophy will have to examine whether the philosophical hypothesis of the existence of the external world is actually presupposed by physical research, whether special existence assumptions are made in physical theories, and whether every theory — in particular relativity and quantum theories — can be cast in purely physical terms. Furthermore, the new realism will have to recognize that, even though physical theory aims at depicting the world in a subject-free fashion, this conceptual reconstruction of things is symbolic, incomplete, and tentative rather than literal, exhaustive, and final. Whether our generation will succeed in cleansing physics from psychological elements and in working out a new critical realistic epistemology, and moreover one favoring research rather than blocking it, remains to be seen. What seems undeniable is that an increasing thirst is being felt for both projects, the scientific and the philosophical one.

The present volume witnesses to the recent surge of interest in a realistic approach. It consists of papers prepared for the International Symposium on the Foundations of Physics, held at Oberwolfach in

July 1966 by the Académie Internationale de Philosophie des Sciences, and sponsored by the International Union for History and Philosophy of Science, the American Institute of Physics, and the Albert-Ludwigs-Universität at Freiburg i. Br. Every contributor to this volume has his own philosophy. Possibly realism, the hypothesis that there is an autonomous external world, is the nonempty intersection of their sets of beliefs. It should suffice to explore a fresh look at physics.

Physikalisches Institut der Universität   MARIO BUNGE
Freiburg i. Br., November 1966

References

[1] EDDINGTON, A. S.: The mathematical theory of relativity, introduction. Cambridge: Cambridge University Press 1923.

[2] CARATHÉODORY, C.: Über die Bestimmung der Energie und der absoluten Temperatur mit Hilfe von reversiblen Prozessen. Sitz. ber. Preuss. Akad. Wiss., Physik.-math. Kl. **1925**, 39.

[3] HEISENBERG, W.: Über den anschaulichen Inhalt der quantentheoretischen Kinematik und Mechanik. Z. Physik **43**, 172 (1927).

[4] BRIDGMAN, P. W.: The logic of modern physics. New York: MacMillan, 1927. Bridgman's later qualification that "verbal" or "pencil and paper" operations were also meant, was paid no attention to — something he deplored in Daedalus **88**, 518 (1959).

[5] FEIGL, H., M. SCRIVEN, and G. MAXWELL (Eds.): Minnesota studies in the philosophy of science. Minneapolis: Minnesota University Press 1956, 1958 and 1962; —, and G. MAXWELL: Current issues in the philosophy of science. New York: Holt, Rinehart & Winston 1961. COLODNY, R. G., and C. G. HEMPEL (Eds.): Frontiers of science and philosophy. Pittsburgh: Pittsburgh University Press 1963. BUNGE M. (Ed.): The critical approach. New York: Free Press 1964.

[6] BOHR, N.: Atomic physics and human knowledge. New York: John Wiley & Sons 1958.

[7] HEISENBERG, W.: Physics and philosophy. New York: Harper & Brothers 1958.

[8] — Cited with approval by N. BOHR: Atomic theory and the description of nature. Cambridge: Cambridge University Press 1934.

[9] WEISSKOPF, V. F.: Quality and quantity in quantum physics. Daedalus **88**, 592 (1959).

Chapter 1

Quantum Mechanics without "The Observer"

Karl R. Popper

Department of Philosophy, L.S.E., University of London, Great Britain

This is an attempt to exorcize the ghost called "consciousness" or "the observer" from quantum mechanics, and to show that quantum mechanics is as "objective" a theory as, say, classical statistical mechanics. My thesis is that the observer, or better, the experimentalist, plays in quantum theory exactly the same role as in classical physics. *His task is to test the theory.*

The opposite view, usually called the *Copenhagen interpretation of quantum mechanics*, is almost universally accepted. In brief it says that *"objective reality has evaporated", and that quantum mechanics does not represent particles, but rather our knowledge, our observations, or our consciousness, of particles.* (Cp. [*28*], p. 100.)

If a mere philosopher like myself opposes a ruling dogma such as this, he must expect not only retaliation, but even derision and contempt. He may well be browbeaten (though I am happy to remember how kindly and patiently I was treated by NIELS BOHR) with the assertion that *all competent physicists know that the Copenhagen interpretation is correct* (since it has been "proved by experiment").

It seems therefore necessary to point out that this assertion is historically mistaken, by referring to physicists who like EINSTEIN, PLANCK, VON LAUE, or SCHRÖDINGER, are as competent as any, and who (unlike EINSTEIN, PLANCK, VON LAUE, and SCHRÖDINGER) were even at one time fully convinced adherents to the Copenhagen interpretation, but who do not now "regard the new interpretation as conclusive or convincing" as HEISENBERG puts it (in [*27*], p. 16).

There is, first, LOUIS DE BROGLIE, a one-time adherent to the Copenhagen interpretation; and his former pupil, JEAN-PIERRE VIGIER.

There is, next, ALFRED LANDÉ, also one of the great founders of quantum theory in the years 1921 to 1924 who later (1937 and 1951) wrote two textbooks on quantum mechanics entirely in the Copenhagen spirit, but who has more recently ([*36, 37, 38*]) become one of the leading opponents of the Copenhagen interpretation.

There is DAVID BOHM who published in 1951 a textbook, *Quantum Theory* [*1*], which was not only orthodox in the Copenhagen sense but one of the clearest and fullest, most penetrating and critical presentations of the Copenhagen point of view ever published. Shortly afterwards, under the influence of EINSTEIN, he tried new ways, and arrived in 1952 [*2*] at a tentative theory (revised in [*2a*]) whose logical consistency proved the falsity of the constantly repeated dogma (due to VON NEUMANN [*46*]) that the quantum theory is *"complete"* in the sense that it must prove incompatible with any more detailed theory.

There is MARIO BUNGE who in 1955 published a paper, "Strife about Complementarity" [*12*].

There is the German physicist, FRITZ BOPP, who explicitly subscribes to the Copenhagen interpretation, in an epistemological paragraph of a most interesting paper; thus he writes, for example, "Naturally our considerations do not mean any alteration of the mathematical concept of complementarity." (Cp. [*8*], pp. 147f.) Yet he develops there (and in previous publications) a theory with which EINSTEIN would hardly have had any quarrel since, on lines not dissimilar to EINSTEIN's (cp. [*19*], pp. 671f.), BOPP interprets the quantum theoretical formalism as an extension of classical statistical mechanics; that is, as a theory of ensembles.

I have given this brief and of course quite incomplete list of dissenters merely to combat the historical myth that only philosophers (and totally incompetent or senile physicists) can doubt the Copenhagen interpretation. But before proceeding to criticize this interpretation in some detail, I should like to discuss two points.

(a) In a very important sense, which to my knowledge has been usually overlooked, the Copenhagen interpretation ceased to exist long ago.

(b) Most physicists who quite honestly believe in it do not pay any attention to it in actual practice.

As to point (a), we must not forget that "the new quantum theory" or "quantum mechanics" was, to start with, and until at least 1935, simply another name for *"the new electromagnetic theory of matter"*.

In order to realize fully how the theory of the atom, and therefore the theory of matter, were identified with the theory of the electromagnetic field, we may for example turn to EINSTEIN, who said in 1920: "... according to our present conceptions the *elementary particles are ... nothing but condensations of the electromagnetic field ....* Our ... view of the universe presents two realities ..., namely, gravitational ether and electromagnetic field, or — as they might also be called — space and matter." (Cp. [*17*], p. 22. The italics are mine.)

Quantum mechanics was regarded by its adherents as *the final form of this electromagnetic theory of matter.* That is to say, the formalism was regarded, first of all, as *the theory of electrons and protons* and thereby as *the theory of the constitution of atoms: of the periodic system of elements and their physical properties: and of the chemical bond, and thus of the physical and chemical properties of matter.*

A very impressive statement of the view held by almost all physicists at least up to the discovery of the positron in 1932 is due to ROBERT A. MILLIKAN:

"Indeed, nothing more beautifully simplifying has ever happened in the history of science than the whole series of discoveries culminating about 1914 which finally brought practically universal acceptance to the theory that the material world contains but two fundamental entities, namely, positive and negative electrons, exactly alike in charge, but differing widely in mass, the positive electron — now usually called a *proton* — being 1850 times heavier than the negative, now usually called simply the *electron*." ([*44*], p. 46; the italics are mine. Cp. also [*43*], p. 377.)

In fact until at least 1935 some of the greatest physicists (cp. EDDINGTON's [*16*]) believed that, with the advent of quantum mechanics, *the electromagnetic theory had entered into its final state,* and that the results of quantum mechanics strongly confirmed that *all matter consisted of electrons and protons.* (Neutrons and neutrinos had also been admitted, somewhat grudgingly, but it was thought that neutrons were protons + electrons; and that neutrinos might not be much more than a mathematical fiction; while positrons were regarded as "holes" in the sea of electrons.)

This theory that matter consists of protons and electrons died long ago. Its ailment (though it first remained hidden) started with the discovery of the neutron and also of the positron (which the Copenhagen authorities refused to believe in at first): and it received its final blow with the discovery of the sharply distinct *levels of interaction*, of which the electromagnetic forces constitute just one among at least four:

1. Nuclear forces.
2. Electromagnetic forces.
3. Weak decay interactions.
4. Gravitational forces.

Moreover, the hope of solving within quantum mechanics such classical problems of the electromagnetic theory as the explanation of the electronic charge has been practically abandoned.

In the light of this situation, we may now look back upon the titanic struggle between EINSTEIN and BOHR. The problem posed by EINSTEIN was whether quantum mechanics was *"complete"*. EINSTEIN said no. (Cp. [*21*].) BOHR said yes.

I have no doubt that EINSTEIN was right. But even today we can read that it was BOHR who won that famous battle. This view persists largely because EINSTEIN's attack upon BOHR's assertion of the *completeness* of quantum mechanics was interpreted by the Copenhagen school as an attack upon *quantum mechanics itself and its "soundness" or consistency*. But this entails that we accept (i) the identification of the Copenhagen interpretation with the quantum theory, and (ii) BOHR's shift of the problem from *completeness* to *soundness* (= *freedom from contradiction*). Yet as EINSTEIN had offered his own (statistical) interpretation of quantum theory, he clearly accepted its consistency.

As to point (b), that is, as to my assertion that most physicists who honestly believe in the Copenhagen interpretation do not pay any attention to it in actual practice, an excellent example is FRITZ BOPP [8], since he believes (as do EINSTEIN, PODOLSKY, and ROSEN) that particles possess both sharp positions and momenta at the same time, while the Copenhagen school believes this to be false, or "meaningless", or "unphysical". To quote a formulation of LANDÉ's of 1951 (before he turned against the Copenhagen interpretation): "The classical idea of particles breaks down under the impact of the uncertainty relations. It is unphysical to accept the idea that there *are* particles possessing definite positions and momenta at any given time, and then to concede that these data can never be confirmed experimentally, as though by a malicious whim of nature." ([39], p. 42. LANDÉ continues by quoting NIELS BOHR [6].) But what I have mainly in mind in connection with my point (b) is this. Admittedly, the formalism of quantum mechanics is still applied by physicists to the old problems, and its methods are, with many modifications, partly used in connection with the many new problems of nuclear theory and elementary particle theory. This is certainly a great credit to its power. Yet at the same time, most *experimentalists*, though much concerned with the limits of precision of their results, do not seem to be more worried about the role of the observer or about interfering with their results than they are in connection with sensitive classical experiments; and most *theorists* are quite clear that a new and much more general theory is needed: they all seem to be in search of *a really revolutionary new theory*.

In spite of all this, it still seems necessary *to discuss the Copenhagen interpretation;* that is, more precisely, the claim that, in atomic theory, we have to regard *"the observer"* or *"the subject"* as particularly important, because atomic theory takes its peculiar character largely from *the interference of the subject or the observer (and his "measuring agencies") with the physical object under investigation*. To quote a typical statement of BOHR's: "Indeed, the *finite interaction between object and measuring agencies* ... entails the necessity of a final renunciation of the classical ideal ... and a radical revision of our attitude towards the problem of physical reality." (Cp. [4], pp. 232f.)

Similarly HEISENBERG: "... the traditional requirement of science ... permits a division of the world into subject and object (observer and observed) .... This assumption is not permissible in atomic physics; the interaction between observer and object causes uncontrollable large changes in the system [that is] being observed, because of the discontinuous changes characteristic of the atomic processes." (Cp. [26], pp. 2f.) Accordingly, HEISENBERG suggests that "it is now profitable to review the fundamental discussion, so important for epistemology, of the difficulty of separating the subjective and the objective aspects of the world". (Cp. [26], p. 65; see also [46], pp. 418—421.)

As opposed to all this I suggest that, in practice, physicists do their measurements and experiments today fundamentally in the same way as they did them before 1925. If there is an important difference, then it is that the degree of indirectness of measurements has increased as well as the degree of "objectivity": where 30 or 40 years ago physicists used to look through a microscope to take a "reading", there are now photographic films, or automatic counters, which do the "reading". And although a photographic film has to be "interpreted" (in the light of a theory), it is in no way physically "interfered with" or "influenced" by this interpretation. Admittedly, many experimental tests have now largely a statistical character, but this makes them no less "objective": their statistical character (often processed automatically by counters and computers) has nothing to do with the alleged intrusion of the observer, or of the subject, or of consciousness, into physics, although the preparation or setting up of an experiment obviously has: *it depends on theory*.

Our *theories* which guide us in setting up our experiments have of course always been our inventions: they are inventions or products of our "consciousness". But that has nothing to do with the scientific status of our theories which depends on factors such as their simplicity, symmetry, and explanatory power, and the way they have stood up to critical discussion and to crucial experimental tests; and on their truth (correspondence to reality), or nearness to truth. (Cp. [49], ch. 10.)

Perhaps this is the best place to insert a few logical remarks on the *distinction between theories and concepts;* remarks which, although what follows does not depend on them, may yet help to remove some obstacles that block the way to a critical understanding of the situation in quantum theory.

What we are seeking, in science, are *true theories* — true statements, true descriptions of certain structural properties of the world we live in. These theories or systems of statements may have their instrumental use;

yet what we are seeking in science is not so much usefulness as *truth; approximation to truth; and understanding.*

Thus theories are described wrongly if they are described as being *nothing but* instruments (for example, instruments of prediction), though they are as a rule, among other things, also useful instruments. But infinitely more important for the scientist than the question of the usefulness of theories is that of their *objective truth*, or their nearness to the truth, and the kind of *understanding* of the world, and of its problems, which they may open up for us. The view that theories are *nothing but* instruments, or calculating devices (cp. [49], chapter 3), has become fashionable among quantum theorists, owing to the Copenhagen doctrine that quantum theory is *intrinsically ununderstandable* because we can understand only *classical "pictures"*, such as "particle pictures" or "wave pictures". I think this is a mistaken and even a vicious doctrine.

Theories are also described quite wrongly as "conceptual systems" or "conceptual frameworks". It is true that we cannot construct theories without using words or, if the term is preferred, "concepts". But it is most important to distinguish between statements and words, and between theories and concepts. And it is important to realize that it is a mistake to think that a theory T_1 is *bound* to use a certain conceptual system C_1: *one* theory T_1 may be formulated in many ways, and may use many different conceptual systems, say C_1 and C_2. Or to put it another way: two theories, T_1 and T_2, should be regarded as one if they are logically equivalent, even though they may use two totally different "conceptual systems" (C_1 and C_2) or are conceived in totally different "conceptual frameworks". I do not happen to believe that SCHRÖDINGER [59] and ECKART [15] have validly established the full logical equivalence of wave mechanics and matrix mechanics: there are some loopholes in these equivalence proofs. In this point I agree with NORWOOD RUSSELL HANSON'S [25] (and E. L. HILL'S [30]), although some of my views on the logic of the equivalence or identity of theories differ somewhat from HANSON'S.

Yet I do not think that such a proof is impossible, *in spite of the great difference between the conceptual frameworks of the two theories.* (What would be needed for a valid proof is something approaching an axiomatization of both theories, and a proof that to every theorem $t_{1,n}$ of T_1 corresponds a theorem $t_{2,n}$ of T_2 such that, with the help of some system of definitions of the concepts of T_1 and of T_2 we can show that $t_{1,n}$ and $t_{2,n}$ are logically equivalent. It would not be necessary for either T_1 or T_2 itself to contain the means needed for formulating these definitions; for these means may be supplied by some *extensions* of the theories. Incidentally, the fact that definitions may be needed for such an equivalence proof does not mean that they are needed within a physical theory.)

Now since theories *can* be equivalent even though their "underlying" conceptual frameworks are utterly different (there are many other examples showing that this may be possible), it is clearly a mistake to identify a theory with its "underlying" conceptual framework or even to believe that these two *must* be very closely related. The conceptual framework of a theory *may* be replaced by a very different one without changing the theory essentially; and *vice versa:* incompatible theories *may* be expressed within the same conceptual framework. (For example, if we replace NEWTON's inverse square law by an inverse law with the power 2.0001, then we have a different theory within the same framework; and the difference will increase if the difference between the two parameters becomes greater. We might even introduce into NEWTON's theory a finite velocity for gravitational interactions and still say that we are operating within the same conceptual framework. If the velocity is very great, the two theories may be experimentally indistinguishable; if it is small, the theories may differ widely in their empirical implications, though still remaining within the same conceptual framework.)

What is of real importance for the pure scientist is the *theory*. And the theory is not merely an "instrument" for him, it is more: he is interested in its truth, or in its approximation to the truth. (Cp. [*49*], chapter 10.) The conceptual system, on the other hand, is exchangeable and is one among several possible instruments that may be used for formulating the theory. It provides merely a language for the theory; perhaps a better and simpler language than another, perhaps not. In any case, it remains (like every language) to some extent vague and ambiguous. It cannot be made "precise": the meaning of concepts cannot, essentially, be laid down by any definition, whether formal, operational, or ostensive. Any attempt to make the meaning of the conceptual system "precise" by way of definitions must lead to an infinite regress, and to merely *apparent* precision, which is the worst form of imprecision because it is the most deceptive form. (This holds even for pure mathematics.)

Thus we are ultimately interested in theories and in their truth, rather than in concepts and their meaning.

This point, however, is rarely seen. HEINRICH HERTZ said (and WITTGENSTEIN repeated it) that in science we make ourselves "*pictures*" ("*Bilder*") of the facts, or of reality; and he said that we choose our "pictures" in such a way that "the logically necessary consequences" ("*die denknotwendigen Folgen*") of the "pictures" agree with "the necessary natural consequences" ("*die naturnotwendigen Folgen*") of the real objects or facts. Here it is left open whether the "pictures" are *theories* or *concepts*. MACH, in discussing HERTZ (cp. [*41*], p. 318), suggested that we should interpret HERTZ's "pictures" as "concepts". BOHR's view seems to be similar when he speaks (as he so often does) of the "*particle picture"* and the *"wave picture"*; in fact, his way of speaking indicates strongly the (at least indirect) influence of HERTZ and MACH.

But "pictures" are unimportant. They are especially unimportant if they are more or less synonymous with "concepts", and almost as unimportant when they are meant to characterize theories. *A theory is not a picture.* It need not be "understood" by way of "visual images": *we understand a theory if we understand the problem which it is designed to solve, and the way in which it solves it better, or worse, than its competitors.* Some people may combine this kind of understanding with visual images, others may not. But the most vivid visualization does not amount to an understanding of a theory unless these other conditions are realized: an understanding of the problem situation, and of the arguments for and against the competing theories.

These considerations are important because of endless talk about the "particle picture" and the "wave picture" and their alleged "duality" or "complementarity", and about the alleged necessity, asserted by BOHR, of using "classical pictures" because of the (admitted but irrelevant) difficulty, or perhaps impossibility, of "visualizing" and thus "understanding" atomic objects. But *this* kind of "understanding" is of little value; and the denial that we can understand quantum theory has had the most appalling repercussions, both on the teaching and on the real understanding of the theory.

In fact, all this talk about pictures has not the slightest bearing on either physics, or physical theories, or the understanding of physical theories. And the fashionable thesis that it is vain to try to "understand" modern physical theories because they are essentially "ununderstandable" (though useful instruments for calculation) amounts to the somewhat absurd assertion that we cannot know what problems they are intended to solve, or why they solve them better, or worse, than their competitors.

If *concepts* are comparatively unimportant, *definitions* must also be unimportant. Thus although I am pleading here for *realism* in physics, I do not intend to define "realism" or "reality". In pleading for realism I wish, in the main, to argue that nothing has changed since GALILEO or NEWTON or FARADAY concerning the status or the role of the "observer" or of our "consciousness" or of our "information" in physics. I am at the same time quite ready to point out that even in NEWTON's physics, "space" was somewhat less real than "matter" (because although it acted upon matter it could not be acted upon); and that in EINSTEIN's special theory of relativity an inertial frame was less real than a spatio-temporal coincidence of two events, or the spatio-temporal distance between them. In a similar way, the number of degrees of freedom of a physical system is a more abstract idea, and perhaps less real, than the atoms or molecules constituting the system; but still, I should be opposed to saying that the degrees of freedom of a system are not real, that they are *nothing but* a conceptual device, and not a real *physical property of the system.* In other words, I do not intend to argue about words, including the word "real"; and by and large I regard as excellent LANDÉ's suggestion to call physically real what is "kickable" (and able to kick back if kicked) — though there are, I am inclined to think, degrees of kickability: we can't kick quasars, DAVID BOHM reminds me.

I have been in doubt whether I should not perhaps first analyse and criticize the central tenets of the Copenhagen interpretation, and then later show that a perfectly realistic interpretation of the theory is possible. I have decided to proceed differently. I am going to expound, in the form of thirteen theses and a summary, my own realistic interpretation, for what it is worth; and I shall criticize the Copenhagen interpretation as I go along. I am sure I shall shock many physicists who, after having reached my fourth, or at the most my sixth thesis, will stop reading this rubbish: it is to help them not to waste their time that I have decided to proceed as I do.

1. My first thesis concerns the most important thing for understanding quantum theory: the kind of *problems* which the theory is supposed to solve. These, I assert, are essentially *statistical problems.* (a) It was so with PLANCK's problem in 1899—1900 which led to his radiation formula. (b) It was so with EINSTEIN's photon hypothesis and his derivation of PLANCK's formula. (c) It was so (at least in part) with BOHR's problem of 1913 which led to his theory of spectral emissions: the explanation of the Rydberg-Ritz combination principle was, clearly, a statistical problem (especially after EINSTEIN's photon hypothesis had been proposed). Admittedly, there was a second problem, thought by BOHR to be *the* fundamental one: the problem of *atomic stability,* or of the "stationary state" of non-radiating electrons in the atom. BOHR "solved" this problem — by a postulate (of "quantum states" or "preferred orbits"). So far as there is any *explanatory* solution to this problem, it is due to wave mechanics; which *in the light of* BORN's *interpretation* means that it is due to the substitution of a statistical problem for a mechanical problem. (See below.) (d) It was so with the set of problems which were solved first by BOHR's most fruitful "principle of correspondence": these were, in the main, problems of the *intensities* of the emitted spectral lines. However, BOHR's correspondence arguments were largely qualitative or, at best, approximations. The central problem which led to the new quantum mechanics was to improve on this by obtaining exact statistical results.

However, this is not at all the way in which BOHR and his school looked at the problem. They did *not* look for a generalization of classical *statistical* mechanics, but rather for a "*generalization of classical [particle] mechanics* suited to allow for the existence of the quantum of action", as BOHR put it as late as 1948; a generalization of particle mechanics which would offer "a frame sufficiently wide to account for ... the characteristic features of *atomic stability which gave the first impetus to the development of quantum mechanics* ...". (Cp. [5], p. 316. The italics are mine.)

Most formulations of the *problem* of quantum mechanics which I have been able to find are similar, except perhaps those "inductivistic" ones that start from the experiments and look upon theory as "the attempt to classify and synthesize the results ... of scientific experiment" (cp.[26], p. 1, and [29]), as if the scientific experiments referred to were not, in the main, only the results of theoretical problems, and significant only because of their conflict with, or support of, some theory. (A similar inductivist attitude appears to be DIRAC's starting point, when he discusses "The Need for a Quantum Theory". (Cp. [14], pp. 1 ff.)

I should admit, however, that BOHR's (in my opinion mistaken) programme of reforming particle mechanics so as to solve the problem of atomic stability appeared to have some prospect of being successfully carried out between 1924 and 1926. I refer, of course, to LOUIS DE BROGLIE's doctoral thesis of 1923—1924 in which he applied to electrons the Einsteinian idea that photons were somehow "associated" with waves, and showed that BOHR's quantized "preferred orbits" (and with them, stability) could be explained by wave interference. This was without doubt one of the boldest, deepest, and most far-reaching ideas in this whole development.

DE BROGLIE's idea was, quite consciously, an inversion of EINSTEIN's idea of associating light quanta or photons with light waves. In EINSTEIN's theory, which thus was the model of DE BROGLIE's, light is emitted and absorbed in the form of "particles" or "light quanta" or "photons"; and thus in the form of things which have a pretty sharp spatio-temporal location, at least while they *interact with matter* by being emitted or absorbed. Light is, however, *propagated* like waves. The square of the amplitude of these waves determines, according to EINSTEIN, the density (that is, the statistical probability) of the photons; and the amplitude of the waves at the place where an atom (in an appropriate state) or a free electron is located determines the probability of the absorption of a photon.

However it was more than two years, during which DE BROGLIE's theory of electrons grew into SCHRÖDINGER's "wave mechanics", before MAX BORN applied to this new wave mechanics the statistical interpretation of the relationship between photons and light waves which we owe to EINSTEIN. MAX BORN himself says about his statistical interpretation of wave mechanics: "The solution ... was suggested by a remark of EINSTEIN's about the connection between the wave theory of light and the photon hypothesis. The intensity [of course, what is meant is the square of the amplitude] of the light waves was to be a measure of the density of the photons or, more precisely, of the probability of photons being present." (Cp. [*10*], p. 104.)

Thus through BORN's statistical interpretation of matter waves even the one problem of quantum theory which appeared not to be statistical — the problem of atomic stability — was reduced to, or replaced by, a statistical problem: BOHR's quantized "preferred orbits" turned out to be those for which the *probability* of an electron's being found on them differed from zero.

All this is to support my thesis that *the problems of the new quantum theory were essentially of a statistical or probabilistic character.*

2. My second thesis is that *statistical questions demand, essentially, statistical answers.* Thus quantum mechanics must be, essentially, a statistical theory.

I believe that this argument (although its validity is by no means generally admitted) is perfectly straightforward and logically cogent. (The argument may be traced back to RICHARD VON MISES [*45*] and it has been beautifully illustrated by ALFRED LANDÉ; cp. [*36*], pp. 3 f., and [*38*], pp. 27 ff. and 39.)

Statistical conclusions cannot be obtained without statistical premises. And therefore answers to statistical questions cannot be obtained without a statistical theory.

Yet largely owing to the fact that the *problems* of the theory were not (and still often are not) seen to be statistical, other reasons were invented to explain the widely admitted statistical character of the theory.

Foremost among these reasons is the argument that it is our (necessary) *lack of knowledge* — especially the limitations to our knowledge discovered by HEISENBERG and formulated in his *"principle of indeterminacy"* or *"principle of uncertainty"* — which forces us to adopt a probabilistic, and consequently a statistical, theory. (This argument is criticized in my fifth thesis below.)

3. My third thesis is that it is this mistaken belief that we have to explain the probabilistic character of quantum theory by our (allegedly necessary) *lack of knowledge,* rather than by the statistical character of our problems, which has led to *the intrusion of the observer, or the subject, into quantum theory.* It has led to this intrusion because the view that a probabilistic theory is the result of lack of knowledge leads inescapably

to the *subjectivist interpretation of probability theory;* that is, to the view that the probability of an event measures the degree of somebody's (incomplete) knowledge of that event, or of his "belief" in it.

However, as I have tried to show for many years, it would be sheer magic if we were able to obtain knowledge — statistical knowledge — out of ignorance. (Cp. [*50, 53, 54, 55*].)

4. My fourth thesis is that, as a consequence, we are faced with what I shall call *the great quantum muddle.* (It seems to me that the only adherent to BOHR's "Principle of Complementarity" who is free of this muddle — following almost exactly EINSTEIN's despised ideas in a new garb — is FRITZ BOPP, in his paper [*8*].)

In order to explain this great muddle, I shall have to say a few words about statistical theories.

Every probabilistic or statistical theory assumes the following.

(a) Certain *events* (5 turning up) which happen to certain *elements* (dice) in certain *experimental situations* (being shaken in a beaker, and thrown on a table). These form the "population" for our statistics.

(b) Certain physical properties of these events, elements, and experimental situations; for example that the dice are of homogeneous material, and that only one of the six sides is marked with a "5"; and that the experimental situation permits a certain width of variation.

(c) A set of the *possible* events (*possible* under the experimental conditions), called the points in the *sample space* or the *probability space* (the notion stems from RICHARD VON MISES).

(d) A number associated with each point (or, in the case of a continuous sample space, with each region) of the sample space, determined by some mathematical function, called the distribution function. (The sum of these numbers is equal to 1; this can be achieved by some "normalization".) In the continuous case the distribution function is a density function.

Example: our sample space may be the United Kingdom, or more precisely, the set of events of a man or a woman living at some spot in the United Kingdom. The distribution function can be given by a (continuous) density distribution (normalized to 1) of the population; that is, the actual number of people living in a region, "normalized" by being divided by the total population of the United Kingdom. We then can say that this *information* helps us to answer all questions of the type: what is the probability that an Englishman lives at a certain spot (region); or that an Englishman lives in "the South of England"? (Here we assume that we have a proper division between North and South.)

Now it is clear that the statistical distribution function (whether normalized or not) may be looked upon as *a property characterizing the sample space* — in our case the United Kingdom. It is *not* a physical property characteristic of the *events* (5 turning up; or of Mr. Henry Smith's, a resident in the United Kingdom, being domiciled in Oxford); still less is it a property of the *elements* (the die; or Mr. Smith).

This is particularly clear of Mr. Smith: he is, for the statistical theory, nothing but an element under consideration. (In fact, the statistical theory will tell us almost the same about Mr. Smith as it tells us, say, about his bed or his wristwatch: the statistical distributions of these physically very different elements will be almost identical.) It is, perhaps, less clear of the die: in this case the distribution function is, we conjecture, *related* to its physical properties (its having six sides, the homogeneity of its material). However, this relation is not as close as it may seem at first sight. For the distribution function will be the same for big or small dice, and for dice made of some light plastic or uranium. And the probability of 5 turning up will be the same for all dice that have only one side marked "5" — whatever the markings of the other sides may be (though these may greatly influence other probabilities); and it will be a different one for all dice having more, or less, than one side marked "5", or for non-homogeneous dice.

Now what I call the great quantum muddle consists in taking a distribution function, i.e. a statistical measure function characterizing some *sample space* (or perhaps some "population" of events), and treating it as *a physical property of the elements of the population*. It *is* a muddle: the sample space has hardly anything to do with the elements.

Unfortunately many people, including physicists, talk as if the distribution function (or its mathematical form) were a property of the *elements* of the population under consideration. They do not discriminate between utterly different categories or types of things, and rely on the very unsafe assumption that "my" probability of living in the South of England is, like "my" age, one of "my" properties — perhaps one of my physical properties.

Now my thesis is that this muddle is widely prevalent in quantum theory, as is shown by those who speak of a "duality of particle and wave" or of "wavicles".

For the so-called "wave" — the ψ-function — may be identified with the mathematical form of a function, $f\left(P, \dfrac{d}{dt} P\right)$, which is a *function of a probabilistic distribution function* P, where $f = \psi = \psi(q, t)$, and $P = |\psi|^2$ is a density distribution function. (See, for example, the footnote 6, with a reference to E. FEENBERG, in H. MEHLBERG's excellent discussion of LANDÉ's views in [42], p. 363.) On the other hand, the *element* in question has the properties of a particle. The wave shape (in

2*

configuration space) of the ψ-function is a kind of accident which poses a problem to probability theory, but which has next to nothing to do with the physical properties of the particles. It is as if I were called a "Gauss-man" or a "non-Gauss-man" in order to indicate that the distribution function of my living in the South of England has a Gaussian or non-Gaussian shape (in an appropriate sample space).

5. My fifth thesis concerns HEISENBERG'S famous formulae:

$$\Delta E \, \Delta t \geqq h, \tag{1}$$

$$\Delta p_x \Delta q_x \geqq h. \tag{2}$$

I assert that these formulae are, beyond all doubt, validly derivable *statistical formulae* of the quantum theory. But I also assert that they have been *habitually misinterpreted* by those quantum theorists who said that these formulae can be interpreted as determining some upper limits to the *precision of our measurements* (or some lower limits to their imprecision).

My thesis is that these formulae set some lower limits to the *statistical dispersion or "scatter"* of the results of sequences of experiments: they are *statistical scatter relations*. They thereby limit the precision of certain individual *predictions*.

But I also assert that *in order to test these scatter relations, we have to be able (and are able) to make measurements which are far more precise than the range or width of the scatter.*

The situation is like this: a statistical theory may tell us something about the distribution or scatter of the population in the environment of industrial towns. In order to *test* it, it will be necessary to fix the places where people live with a precision far exceeding the range of the predicted scatter. Our statistical laws may tell us that we cannot reduce the scatter below a certain limit. But to conclude from this that we are unable to "measure" the positions of the places where the people live more precisely than the minimum statistical scatter is simply a muddle.

Since HEISENBERG'S formulae in their various proper interpretations are (as will be shown in detail in my next thesis) *statistical laws of nature, derivable from a statistical theory*, it is quite obvious that it is impossible to use them in order to explain why quantum mechanics is probabilistic or statistical. Moreover, being statistical laws, they *add* to our knowledge: it is a mistake to think that they set limits to our knowledge. What they do set limits to is the scatter of particles (or more precisely, the scatter of the result of sequences of certain experiments with particles). This scatter, they tell us, cannot be suppressed. It is also a mistake to think that the alleged limitation to our knowledge could ever be validly used for explaining the statistical character of the quantum theory. (See my eighth thesis, below.) And ultimately, it is just the old muddle again if it is said that the Heisenberg formulae provide us with that *vagueness* which is allegedly needed for asserting without inconsistency the "dual" character of particles and waves; that is, their character as "wavicles".

6. My sixth thesis is that, however remarkable the statistical laws of the theory are, including the Heisenberg formulae (1) and (2), they refer to a population of *particles* (or of experiments with *particles*) which are, quite properly, endowed with positions and momenta (and mass-energy, and various other physical properties such as spin). It is true that the scatter relations tell us that *we cannot prepare experiments* such that we can avoid, upon repetition of the experiment, (1) scattering of the energy if we *arrange* for a narrow time limit, and (2) scattering of the momentum if we *arrange* for a narrowly limited position. But this means only that there are limits to the *statistical homogeneity* of our experimental results. Yet not only is it possible to *measure* energy *and* time, or momentum *and* position, with a precision greater than formulae (1) and (2) seem to permit, but *these measurements are necessary for testing the scatter* predicted by these very formulae.

I shall now try to produce some arguments for what I have said in my last two theses. These arguments will show, incidentally, that the Heisenberg formulae (1) and (2) can be derived from theories which are much older than the commutation relations of quantum mechanics.

We can derive HEISENBERG'S formula

$$\Delta E\, \Delta t \geqq h \tag{1}$$

from PLANCK'S quantum condition of 1900,

$$E = h\,\nu.$$

This leads, in view of the constancy of h, at once to

$$\Delta E = h\,\Delta \nu,$$

a formula in which "Δ" may be interpreted in various ways. In order to obtain HEISENBERG'S formula (1) we only have to combine this formula with an even older principle of optics, the *principle of harmonic resolving power*. (Both HEISENBERG and BOHR base their derivations of the indeterminacy relations directly or indirectly upon this principle; cp. [*26*], pp. 21 and 27.) This principle states that if a monochromatic wave train of frequency ν is cut up by a *time shutter* into one stretch or several stretches ("wave packets") of the duration Δt, then the width $\Delta \nu$ of the spectral line will become

$$\Delta \nu \geqq 1/\Delta t.$$

This is, for various reasons, a remarkable law. (It contains the principle of superposition.) It leads from

$$\Delta E = h \, \Delta \nu$$

immediately to

$$\Delta E \geqq h/\Delta t$$

and thus to formula (1).

But in so deriving formula (1), *we are no longer free* to interpret "Δ" in various ways (for example, as the width of imprecision of a measurement). We are, rather, bound in our interpretation by the meaning given to "Δ" by the principle of harmonic resolving power. This principle interprets "$\Delta \nu$" as the width of spectral lines. Accordingly, PLANCK's principle (in EINSTEIN's interpretation) forces us to interpret this width as the scatter of the energy of the particles (photons) which make up the spectral lines; for a spectral line of frequency ν is to be interpreted as the statistical result of incoming photons of energy $E = h\nu$, and consequently the width $\Delta \nu$ of the spectral line as the range ΔE of the *statistical scatter* of the energies of the photons which together form the spectral line. Thus formula (1) states the law that, if we vary at will the period Δt of our shutter, we are bound to influence inversely the scatter ΔE of the energy of the incoming photons.

This derivation shows clearly that (1) is a *statistical law*, and part of the statistical theory. It can be tested only by ascertaining the distribution of the incoming photons on the photographic film or plate; and in order to do this, we must measure the places where the photons hit the spectral line with an imprecision, say δE, very much smaller than the width ΔE of the line:

$$\delta E \ll \Delta E.$$

Thus the testing of the law expressed by (1) and of its statistical predictions demand that we can *measure* the incoming particle with a precision δE which satisfies

$$\delta E \, \Delta t \ll h.$$

This kind of thing is done every day; and it shows that the Heisenberg formulae are valid for *statistical predictions about many particles, or about sequences of many experiments* with individual particles, but that they are misinterpreted as limiting the precision of *measurements* of individual particles.

There is a derivation of the second Heisenberg formula

$$\Delta p_x \Delta q_x \geqq h \tag{2}$$

which is analogous to the derivation with the help of the *time shutter*. We start again with a (flat) monochromatic wave train ν and cut it;

this time by a *screen* (vertical to the direction z of the beam), *with one slit* of variable width Δq_x. (The *"one slit experiment"*.) When the slit is very wide, there will be only a marginal effect upon the wave train. But when it narrows, we get a scattering (diffraction) effect: the narrower the slit Δq_x, the wider will be the angle by which the rays diverge from their original direction: here another form of the principle of harmonic resolving power applies ($\tilde{v}_x$ is the projection on the x-axis of the wave number, that is, the number of waves per centimetre):

$$\Delta \tilde{v}_x \approx 1/\Delta q_x .$$

Multiplying both sides by h we get

$$h\Delta \tilde{v}_x \approx h/\Delta q_x .$$

Using instead of PLANCK's formula $E = h\nu$ that of DE BROGLIE in the form $p_x = h\tilde{v}_x$, we can write "Δp_x" for "$h\Delta \tilde{v}_x$"; and so we arrive at (2).

When the slit Δq_x is very small we obtain, according to HUYGENS's principle, waves emerging from it which spread not only in the z direction but also in the $+x$ and $-x$ direction (cylinder waves). This means that the particles which, before reaching the slit, had a momentum $p_x = 0$ (since they were proceeding in the z direction), will now have a considerable scatter of momenta Δp_x, in the $+x$ and $-x$ direction. We can test this scatter again by measuring the various momenta with a spectrograph in various positions. There is, in principle, hardly a limit to the precision δp of the measurements of the various momenta in the various directions; that is, we have again

$$\delta p_x \ll \Delta p_x$$

and thus

$$\delta p_x \Delta q_x \ll h .$$

Again, we could not test the statistical law (2) in this *one slit experiment* without these more precise measurements $\delta p_x \ll \Delta p_x$.

Incidentally, *we measure the momentum, p_x,* of the incoming particle *by its position* on the film of the spectrograph. And this is typical. It should hardly be necessary to stress that we almost always measure momenta by positions. (For example, if we measure a Doppler effect, we do so with the help of a spectral line, that is by measuring the position of the line on a photographic plate.) It has, unfortunately, become necessary to emphasize this point, because of BOHR's repeated assertion that position measurements and momentum measurements are incompatible (and "complementary") owing to "the mutual exclusion of any two experimental procedures, permitting the unambiguous definition of complementary physical qualities ...". (Cp. BOHR in [4], p. 234, quoting from his reply [3] to EINSTEIN, PODOLSKY, and ROSEN [21].) The two experimental procedures, we are told by BOHR, exclude each other because the momentum measurements need a *movable screen* (as depicted in [*4*], p. 220), while the position measurement needs a *fixed screen* or a fixed photographic plate. But we often measure momenta by fixed photographic plates, that is, by positions; but never by a movable screen. (Incidentally, the use of BOHR's movable screen would entail at least two *position measurements* of the screen.)

A famous problem for particle theory is posed by the *two slit experiment* (or *n* slit experiment), with two (or more) slits with the (periodic) distance Δq_x between them. This has been recently cleared up by ALFRED LANDÉ ([*38*], pp. 9—12), using the Duane-Landé space-periodicity formula:

$$\Delta p_x = n\,h/\Delta q_x \qquad\qquad (n = 1, 2, \ldots).$$

The two slit experiment turns out to be a space-periodicity experiment with the periodicity Δq_x. The particles transfer to the screen (or the grid) a momentum packet Δp, or its multiples, such that

$$\Delta p_x \Delta q_x = h.$$

As a consequence (as shown by LANDÉ, *loc. cit.*) we get the wave-like fringes.

The usual question "how does the particle which goes through slit 1 'know' that slit 2 is open rather than closed?" can now be reasonably well cleared up. It is the *screen* (or the grid, or the crystal) which "knows" whether there is a periodicity Δq_x built into it or not, and which therefore "knows" whether it can absorb momentum packets of the size $\Delta p_x = h/\Delta q_x$. The particle does not need to "know" anything: it simply interacts with the screen (which "knows"), according to the laws of conservation of momentum *and* of space periodicity; or more precisely, it interacts with the total experimental arrangements (see my eighth and especially my tenth thesis below).

I have so far spoken mainly about particles and their (indirect) measurements, for example, momentum measurements by way of position measurements. But there are other methods, of course: Geiger counters may measure (not very precisely) position, *and* time; and so may Wilson chambers. And the position measurement in a Wilson chamber may be an indirect momentum measurement. However, the time measurement of the incoming particle may be of particular interest to us in every case in which the frequency (or energy) of the emission is very sharp — as it is in the classical case of a BOHR hydrogen atom.

Here we have RYDBERG's constant R, a wave number, so that Rc is a constant frequency, ν_R, which can be calculated, according to HAAS (1910) and BOHR (1913), with great precision from the constants of the theory (μ is the mass of the electron, e its charge):

$$\nu_R = Rc = 2\,\pi^2 e^4 \mu/h^3.$$

Then the Rydberg-Ritz combination principle (formulated by RITZ, using RYDBERG's constant, in 1908, five years before BOHR's theory of the hydrogen atom) asserts for the frequencies, $\nu_{m,n}$, of emission or absorption the relation

$$\nu_{m,n} = \nu_R/m^2 - \nu_R/n^2 \qquad (m,\, n = 1,\, 2,\, \ldots).$$

Multiplied by h this becomes BOHR's quantization rule of emission and absorption (1913). Thus the permissible frequencies $\nu_{m,n}$ and the various corresponding Bohr-energies of the particles can be calculated from first principles, as it were — they are variables which can take on only certain discrete values ("*eigenvalues*" which might be described as quasi-constants). Accordingly, $\Delta\nu_{m,n}$ may be extremely small, and Δt, calculated with the help of the principle of harmonic resolving power, will be large.

But these sharp spectral lines, although they must not be interfered with by means of a time shutter, can be statistically investigated by timing the arrivals of the photons (which gives also the time of emission) by means such as a Wilson chamber or a Geiger counter. (Especially impressive here are the Compton-Simon photographs of high frequency X-ray photons of very precise frequency or energy.) For these arrival times we may get $\delta t \ll \Delta t$, and thus

$$\Delta E\ \delta t \ll h.$$

7. My seventh thesis is that all this, or most of it, was in effect admitted by HEISENBERG.

First I would repeat that the *predictions* of the theory are statistical, with a scatter given by the Heisenberg formulae. The *measurements* which must be *more precise than the scatter* (as I have pointed out) may serve as *tests* of these predictions: *these measurements are retrodictions.*

HEISENBERG saw, and said, that these highly precise retrodictive measurements were possible. *What he did not see was that they had a function in the theory — that they were needed for testing it* (and that they could be tested in their turn).

Thus he suggested, half-heartedly but pretty strongly, that these retrodictive measurements were *meaningless*. And this suggestion was taken up and *turned into a dogma* by the adherents of the Copenhagen interpretation, especially when it was found that there were no vectors in Hilbert space corresponding to any measurements sharper than the formulae (1) and (2).

But this fact does not really create any difficulty. *The vectors in Hilbert space correspond to the statistical assertions of the statistical theory.*

They say nothing about measurements, or about the *tests* of the statistical assertions by the determination of the position and momentum, or of the energy and time, of individual particles.

I shall now quote some evidence for my thesis regarding HEISENBERG's admission that the measurements I have described can be made, and his suggestion that they are, if not completely meaningless, at best pointless and uninteresting, because they merely refer to the past. He says that measurements which "can never be used as initial conditions in any calculation of the future progress of the electron and [which] thus cannot be subjected to experimental verification" are devoid of physical significance. (Cp. [*26*], p. 20.) But this is a double mistake. For (a) the preparation of initial conditions admittedly is very important, but so are test statements which *always* look into the past and whose *main* function is not to be "verifiable" (that is, testable) in their turn but to "verify" (or more precisely to test). And (b) it is a mistake to think that these test statements, though looking into the past, are not "verifiable" (or more precisely, testable) in their turn. On the contrary, it is one of the principles of the quantum theory that every measurement can be "verified" or tested in the sense that its immediate repetition will yield the same result. (This principle, whose author seems to be VON NEUMANN, is not generally valid, unless it is trivially so in the sense explained below, under the heading of my ninth thesis.) Thus to say that these measurements which look into the past "cannot be subjected to experimental verification" is simply mistaken.

In order to show quite definitely that HEISENBERG and I are talking about the same measurements, and that we are in agreement that they are not subject to the uncertainty relations, I wish to remind the reader of the one slit experiment and of the fact that the measurements of p_x with the help of spectrographs at various positions (or of a photographic plate parallel to the horizontal screen) are, in fact, *position* measurements, so that we obtain our total information about position $+$ momentum by way of *two position measurements:* the first is provided by the slit Δq_x, the second by the impact of the particle on the photographic plate. (We can take the frequency — or energy — of the beam as known.) Now it is precisely about such an arrangement consisting of *two position measurements* (which allow us to calculate the position *and* momentum *after* the first and *before* the second measurement) that HEISENBERG says the following:

"The ... most fundamental method of measuring velocity [or momentum] depends on the determination of position at two different times ... it is possible to determine *with any desired degree of accuracy* the velocity [or momentum] before the second [measurement] was made; but it is the velocity *after* this measurement which alone is of importance to the physicist ..." (Cp. [*26*], p. 25. The italics are mine, and I have changed the position of a phrase to improve the readability by avoiding an ambiguity.)

HEISENBERG is even more emphatic concerning experiments in which we measure the position of a particle whose momentum is known (say, because the particle belongs to a monochromatic beam): "... the uncertainty relation does not refer to the past", he writes; "if the velocity of the electron is at first known and the position then exactly measured, the position for times previous to the measurement may be calculated. Then for these past times $\Delta p \, \Delta q$ is smaller than the usual limiting value." (Cp. [*26*], p. 20.) So far we can agree. But now comes our subtle but important disagreement; for HEISENBERG continues: "but this knowledge of the past is of a purely speculative character, since it can never ... be used as an initial condition in any calculation of the future progress of the electron" (this I believe to be true) "and thus cannot be subjected to experimental verification" (this is false, as I shall show).

HEISENBERG adds to this: "It is a matter of personal belief whether such a calculation concerning the past history of the electron can be ascribed any physical reality or not." (*Loc. cit.*)

Almost every physicist who read HEISENBERG opted for "not".

But it is not a matter of personal belief: the measurements in question are *needed* for testing the statistical laws (1) and (2); that is, the scatter relations.

The particular case, of a position measurement of a particle from which retrodictively "the positions for times previous to the measurement may be calculated", as HEISENBERG puts it, plays a *most* important role in physics: if we measure the position of a particle (a photon or an electron) on the photographic film of any spectrograph, then we use this position measurement (together with the known arrangement of the experiment) for calculating, with the help of the theory, the frequency or energy and thus the momentum of the particle; always, of course, retrodictively. To question whether the so ascertained "past history of the electron can be ascribed any physical reality or not" is to question the significance of an indispensable standard method of measurement (retrodictive, of course); indispensable, especially, for quantum physics.

But once we ascribe physical reality to measurements for which, as HEISENBERG admits, $\Delta p \, \Delta q \ll h$, the whole situation changes completely: for now there can be no question whether, according to the quantum theory, *an electron can "have" a precise position and momentum. It can.*

But it was just this fact that was constantly denied: although HEISENBERG made it "a matter of personal belief", BOHR and the Copenhagen interpretation (partly because of the non-existence of those vectors in

Hilbert space) *insisted that an electron just cannot have a sharp position and momentum at the same time.* This dogma is the core of BOHR's thesis that quantum theory is "complete", presumably in the sense that a particle cannot have properties which the theory (allegedly) does not allow to be "measured".

Thus the so-called "paradox" of EINSTEIN, PODOLSKY, and ROSEN (cp. [*21*] and [*3*]) is not a paradox but a valid argument, for it established just this: that we must ascribe to particles a precise position *and* momentum, which was denied by BOHR and his school (though it is admitted by BOPP).

The Einstein-Podolsky-Rosen thought experiment has since become a *real* experiment, in connection with pair-creation, and pair-destruction with photon-pair creation. The times and energies of the pairs can be in principle measured with any degree of precision. Of course, the measurements are *retrodictive:* they are *tests* of the theory. (See for example O. R. FRISCH's [*23*].)

Why did BOHR and his followers deny that $\delta p_x \delta q_x \ll h$ is possible? Because of the great quantum muddle, the alleged dualism of particle and wave: it is said that there are two *"pictures"*, the *particle picture* and the *wave picture*, and that they have been shown to be equivalent or "complementary"; that is to say, both valid. But this "complementarity" or "duality" must break down, it is said, if we allow the particle to have at the same time a sharp position and momentum.

It is from here, and from the subjective interpretation of probability to which we shall turn next, that the subjectivist interpretation of the quantum theory arose — almost of necessity.

8. My eighth thesis results from an attempt of mine to explain, though not to excuse, the great quantum muddle, as I have called it. My thesis is that *the interpretation of the formalism of quantum mechanics is closely related to the interpretation of the calculus of probability.*

By the calculus of probability I mean a *formal calculus* which contains *formal laws* such as

$$0 \leq p(a, b) \leq 1$$

where "$p(a, b)$" may be read "the probability of a relative to b" (or "the probability of a given b").

What "probability" (the function of functor "p") *means*, and what the arguments "a" and "b" *stand for*, is left open to *interpretation.*

It is assumed, however, that there is a set of entities, S, say, to which the arguments $a, b, c, \ldots$, belong; and that if a belongs to S, then $-a$ (read "non-a") also belongs to S; and that if a and b belong to S, then ab (read "a-and-b") also does. Moreover, it is assumed that the meaning of all these symbols, though open to many different interpretations, is partly fixed by a number of formal rules which connect these symbols.

The following formulae are trivial examples of such formal rules:

$$p(a, a) = 1.$$

$$p(a, b) + p(-a, b) = 1, \quad \text{unless } p(-b, b) \neq 0.$$

$$p(a, b) = p(aa, b) = p(a, bb).$$

$$p(a, c) \geq p(ab, c) \leq p(b, c).$$

We may also give a definition of "*absolute probability*", $p(a)$, in terms of "*relative probability*", $p(a, b)$:

$$p(a) = p\left(a, -((-a)a)\right).$$

The task of selecting a number of these *formal rules* so that all the others are derivable from them, is the task of finding one or more suitable *axiomatizations* of the formal calculus of probability. (Cp. [50] and [51].) I mention it only in order to contrast it with the task of finding one or more suitable *interpretations*. (Cp. [53], [54], and [55].)

There is a great variety of interpretations, which may be divided into two main groups: the *subjective* and the *objective* interpretations.

The *subjective interpretations* are those which interpret the number $p(a, b)$ as measuring something like our knowledge, or our belief, in (the assertion) a, given (the information) b. Thus the arguments of the p-function, that is, $a, b, c, \ldots$ are in this case to be interpreted as items of belief or doubt, or items of information, or propositions, or assertions, or statements, or hypotheses.

For a long time it was thought (and it still is thought by many eminent mathematicians and physicists) that we may start from a subjectively interpreted system of probabilistic premises and then *derive from these subjectivist premises statistical conclusions*. However, *this is a grave logical blunder*.

The blunder may be traced back to some of the great founders of probability theory, to JACOB BERNOULLI and especially to SIMÉON DENIS POISSON, who thought that they had discovered, in their derivations of the various forms of the *law of great numbers*, a kind of logico-mathematical *bridge* leading from non-statistical assumptions to statistical conclusions; that is, to conclusions concerning the *frequency* of certain events.

The logical mistake was carefully analysed by RICHARD VON MISES (see especially [45]) and also by myself. (Cp. [50], chapter VIII, and [53].) MISES showed that at some stage or other in the derivation, the non-statistical meaning of the symbols is dropped and tacitly replaced by a statistical one. This is usually done by interpreting a probability approaching 1 as "almost certain" in the sense of "almost always to happen", instead of "almost certain" in the sense of "very strongly believed in" or perhaps "almost known". Sometimes the mistake consists in replacing "almost certainly known" by "known almost certainly to occur". However this may be, the mistake is very clear: from premises about degrees of belief we can never get a conclusion about the frequency of events.

It is strange that this idea that we can derive statistical conclusions from premises expressing uncertainty is still so strong among quantum theorists; for JOHN VON NEUMANN, one of the most influential among them, accepted in his famous book, *Mathematical Foundations of Quantum Mechanics*, the theory of probability of VON MISES. (Cp. [*46*], p. 298, note 156.) Yet VON NEUMANN's praise of this theory does not seem to have induced quantum theorists to study carefully VON MISES's arguments against the existence of a *"bridge"* from non-statistical premises to statistical conclusions.

I do not wish to imply that I accept VON MISES's theory as a whole; but I believe that his criticism of the alleged *"bridge"* from non-statistical premises to statistical conclusions is unanswerable; and I do not even know of any serious attempt to refute it. Nevertheless, the subjective theory, under the name of "Bayesian probability", is widely and uncritically accepted.

I now proceed to the *objective interpretations* of the probability calculus. I shall here distinguish between *three* such interpretations:

(a) *The classical interpretation* (DE MOIVRE, LAPLACE) which takes $p(a, b)$ to be the proportion of equally possible cases compatible with the event b which are also favourable to the event a. For example, let a be the event "at the next throw of this die 5 will turn up" and take b to be the assumption "6 will *not* turn up" (or "only throws other than 6 will be considered as throws"); then $p(a, b) = \frac{1}{5}$.

(b) *The frequency interpretation* or statistical interpretation (JOHN VENN, GEORG HELM, VON MISES) which takes $p(a, b)$ as the relative frequency of the events a among the events b. This interpretation, which I developed by trying to remove some of its difficulties (cp. [*50*], chapter VIII and new Appendix* VI), is one which I upheld for about twenty years (from approximately 1930 to 1950), though I always stressed the existence of other interpretations (cp. [*50*]).

(c) *The propensity interpretation* which I developed from a criticism of my own form of the frequency interpretation and which may at the same time be regarded as a refinement of the classical interpretation.

I shall have to say a few things about each of the three objective interpretations.

In favour of (a), the classical interpretation, it may be said that it is used, almost as a matter of course and obviously with good reason, in situations where we conjecture that we have before us something like *"equally possible cases"*: we do not need to experiment with a regular polyhedron in order to conjecture that, if it is of homogeneous material and has n sides, the probability for each of these sides turning up in any one throw will be $1/n$.

On the other hand, the classical interpretation has been severely criticized on several counts, of which I will mention only two: as it stands it is inapplicable to anything like unequally possible cases such as playing with a loaded die; and it succumbs, like the subjective interpretation, to VON MISES's criticism: there is no logical or mathematical *bridge* (like the law of great numbers) which leads from premises about *possibilities* to statistical conclusions about *relative frequencies.* (MISES showed this in great detail for POISSON's derivation of his law of great numbers.) Nor does it make much sense to say of a ratio of many favourable to many possible cases that, even if it approaches 1, it tells us what is almost certainly going to happen: obviously, there occurs here (as VON MISES stressed) a shift of meaning in proceeding from possibilistic premises to statistical conclusion.

As to (b), the frequency interpretation, I feel confident that I have succeeded (cp. [*50*]) in purging it of all those allegedly unsolved problems which some outstanding philosophers like WILLIAM KNEALE (cp. [*33*]) have seen in it. Nevertheless, I found that a further reform was needed, and I tried to respond to this need in two papers. (Cp. [*54*] and [*55*].)

Thus I come to (c), to the *propensity interpretation* of probability.

Let me first make clear that nothing is further from my mind than an attempt to solve the pseudo-problem of giving a definition of the meaning of probability. It is obvious that the word "probability" can be used perfectly properly and legitimately in dozens of senses, many of which, incidentally, are incompatible with the formal calculus of probability. (For such senses see [*60*] and [*24*].) I do not even wish to say that the propensity interpretation of probability is the best interpretation of the formal probability calculus. I only wish to say that it is the best interpretation known to me *for the application of the probability calculus to a certain type of "repeatable experiment"*; in physics, more especially, and also, I suppose, in related fields such as experimental biology.

I fully agree with those who have criticized the propensity interpretation because they felt it was not clear how to apply it to the betting situation in horse racing. The formal probability calculus is applicable to a large class of "games of chance"; but I do not know how one could apply it to betting on horses. Yet should it be possible to apply it to this kind of betting I should see no reason to fear that the propensity interpretation would not fit this case. In brief, I am not trying to propose universally satisfactory meanings for the words "probable" and "probability", or even a universally applicable interpretation of the formal calculus. But I am trying to propose an interpretation of the probability calculus which is not *ad hoc*, and which solves some of the problems of the *interpretation of quantum theory*.

I shall here explain the propensity interpretation as a development of the classical interpretation. The latter, it will be remembered, explains $p(a, b)$ as the proportion of the equally possible cases compatible with b which are also favourable to the event a.

I propose, as a first step, to omit the word "equally" and to introduce "weights" and thus to speak, instead of "numbers of cases", of the "sum of the weights of the cases". And I propose, as a second step, to interpret these "weights" of the possibilities (or of the possible cases) as *measures of the propensity, or tendency, of a possibility to realize itself upon repetition*.

The main idea of this interpretation can also be put as follows: I propose to distinguish *probability statements* from *statistical statements*, and to look upon probability statements as statements about frequencies in *virtual* (infinite) sequences of well characterized experiments, and upon *statistical statements* as statements about frequencies in *actual* (finite) sequences of such experiments. In probability statements, the "weights" attached to the possibilities are measures of these (conjectural) virtual frequencies, *to be tested by actual statistical frequencies*.

To use an example: if we have a large die containing a piece of lead whose position is adjustable, we may *conjecture* (for reasons of symmetry) that the weights (that is, the propensities) of the six possibilities are *equal* as long as the centre of gravity is kept equidistant from the six sides, and that they become *unequal* if we shift the centre of gravity from this position. For example, we may increase the weight of the possibility of 6 turning up by moving the centre of gravity away from the side showing the figure "6". And we may here interpret the word "weight" to mean "a measure of the propensity or tendency to turn up upon repetition of the experiment". More precisely, we may agree to take as our measure of that propensity the (virtual) *relative frequency* with which the side turns up in a (virtual, and virtually infinite) sequence of repetitions of the experiment.

We then may test our conjecture by a sequence of repetitions of the experiment.

In proposing the propensity interpretation I propose to look upon *probability statements* as statements about some measure of a property (a physical property, comparable to symmetry or asymmetry) of *the whole experimental arrangement;* a measure, more precisely, of a *virtual frequency;* and I propose to look upon the corresponding *statistical statements* as statements about the corresponding *actual frequency.*

In this way we easily get over the objection raised by VON MISES against the classical interpretation, simply by replacing mere *possibilities* by *propensities* which we interpret as tendencies to produce frequencies.

Two further points are very important:

First, the probability is taken to be a property of the *single experiment*, relative to some rule specifying the conditions for accepting another experiment as a *repetition of the first.* For example, in dicing, the minimum time taken in shaking the beaker may or may not form part of this rule, or of these conditions or specifications.

Secondly, we can look upon probability as a *real physical property of the single physical experiment* or, more precisely, of the *experimental conditions* laid down by the rule that defines the conditions for the (virtual) *repetition* of the experiment.

A propensity is thus a somewhat abstract kind of physical property; nevertheless it is a real physical property. To use LANDÉ's terminology, *it can be kicked, and it can kick back.*

Take for example an ordinary symmetrical pin board, so constructed that if we let a number of little balls roll down, they will (ideally) form a normal distribution curve. This curve will represent the *probability distribution* for each single experiment with each single ball of reaching a certain possible resting place.

Now let us "kick" this board; say, by slightly lifting its left side. Then we also kick the propensity, and the probability distribution, since it will become more probable that any single ball will reach a point towards the right end of the bottom of the board. And the propensity will kick back: it will produce a differently shaped curve formed by the balls if we let them roll down and accumulate.

Or let us, instead, remove *one pin.* This will alter the probability for every single experiment with every single ball, *whether or not the ball actually comes near the place from which we removed the pin.* (This has its similarity with the two slit experiment, even though we have here no superposition of amplitudes; for we may ask: "How can the ball 'know' that a pin has been removed if it never comes near the place?" The answer is: the ball does not "know"; but the board as a whole "knows", and changes the probability distribution, or the *propensity,* for *every* ball; a fact that can be tested by statistical tests.)

Thus we can "kick" the probability field by making certain (gradual) changes in the conditions of the experiment, and the field "kicks back" by changing the propensities, an effect which we can test statistically by repeating the experiment under the changed conditions.

But there are further important aspects of the propensity interpretation, which we can again illustrate with the help of the pin board.

We can leave the pin board in its ordinary (symmetrical) state; and we can ask for the probability distribution of reaching the various final positions *for those balls which actually hit a certain pin* (or, alternatively, which hit the pin and then pass on its left side).

This new distribution will be, of course, quite different from the original distribution. It can be calculated from first principles (given a symmetrical board); and we can *test* our calculations in various ways. For example, we can let the balls roll down as usual, but list separately the final positions of those balls that hit the selected pin (or that hit it and pass on its left); or else, we can *remove* those balls at once which do not satisfy this new condition. In the first case, we merely *take note* of the new "position measurement" of the ball; in the second case, we *select* the balls which pass through some predetermined position.

In both cases we shall get tests of the calculated new distribution: the distribution of those balls which have undergone a "position measurement".

The theory of the pin board allows us, of course, to calculate from first principles the new distributions for any pin we choose; in fact, all these new distributions are implicit in calculating the original normal distribution. For this calculation assumes that the ball will hit, with such and such a probability, such and such a pin.

9. Ninth thesis. In the case of the pin board, the transition from the original distribution to one which assumes a "position measurement" (whether an actual one or a feigned one) is not merely analogous, *but identical with the famous "reduction of the wave packet"*. Accordingly, this is not an effect characteristic of quantum theory but of probability theory in general. (Cp. [50], section 76.)

Take our pin board example again: given not only the topography of the board but also its inclination and a few more facts, we may look at the probability distribution as a kind of descending wave front, starting to descend when the particle enters the board through its slit Δq. There will be no interference of amplitudes: if we have two slits Δq_1 and Δq_2, the two probabilities themselves (rather than their amplitudes) are to be added and normalized: we cannot imitate the two slit experiment. But this is not our problem at this stage. What I wish to show is this: we may calculate a probability wave, descending to the bottom of the board, and forming there a normal distribution curve very much like a wave packet.

Now if we let *one* actual ball roll down, then we can look at it from various points of view.

(a) We may say that the experiment as a whole determines a certain probability distribution and retains it (upon repetition) irrespective of the particular pins hit by the ball.

(b) We may say that every time the ball actually hits a certain pin (or, say, passes on its left side), the *objective probability* distribution (the propensity distribution) is "suddenly" changed, whether or not anybody takes note of the course of the ball. But this is merely a loose way of saying the following: *if we replace the specification of our experiment by another one* which specifies that the ball hits that particular pin (or passes on its left), then we have a different experiment and accordingly get a different probability distribution.

(c) We may say that the *knowledge*, or the *information*, or the *consciousness*, or the *realization*, that a position measurement has taken place, leads to the *"collapse" or "reduction" of the original wave packet* and to its replacement by a new wave packet. But in speaking in this way, we only say the same as we said before under (b); except that we now use subjectivist language (or a subjectivist philosophy).

Obviously, if we do not, *know* which pin the ball has hit, we do not *know* with which new experimental set of conditions (propensities) we could replace, in this particular case, the old ones. But whether we know this or not — we *did* know from the very start that there was such and such a probability of the ball hitting such and such a pin, and thereby changing its propensity of hitting other pins, and ultimately of reaching a certain point (or column), a, at the bottom of the board. It was on this knowledge that we based our calculation of the original probability distribution (wave packet).

Let us call our original specification of the pin board experiment "e_1" and let us call the new specification (according to which we consider or select only those balls which have hit a certain pin, q_2, say, as repetitions of the new experiment) "e_2". Then it is obvious that the two probabilities of landing at a, $p(a, e_1)$ and $p(a, e_2)$, will not in general be equal, because the two experiments described by e_1 and e_2 are not the same. But this does not mean that the new information which tells us that the conditions e_2 are realized in any way changes $p(a, e_1)$: from the very beginning we could calculate $p(a, e_1)$ for the various a's, and also $p(a, e_2)$; and we knew that

$$p(a, e_1) \neq p(a, e_2).$$

Nothing has changed if we are informed that the ball has actually hit the pin q_2, except that we are now free, *if we so wish*, to apply $p(a, e_2)$ *to this case;* or in other words, we are free to look upon the case as an instance of the experiment e_2 instead of the experiment e_1. But we can, of course, continue to look upon it as an instance of the experiment e_1, and thus continue to work with $p(a, e_1)$: the probabilities (and also the probability packets, that is, the distribution for the various a's) are *relative probabilities:* they are *relative to what we are going to regard as a repetition of our experiment;* or in other words, they are relative to what experiments are, or are not, *regarded as relevant to our statistical test.*

Take another example, a very famous one, due to EINSTEIN, and discussed by HEISENBERG ([26], p. 39) and by myself ([50], end of section 76; English edition, pp. 235 f.). Take a semi-transparent mirror, and assume that the probability that light will be reflected by it is $\frac{1}{2}$. Thus the probability that light will pass through will also be $\frac{1}{2}$, and we have, if the event "passing through" or "transmitted" is a, and the experimental arrangement b,

$$p(a, b) = \tfrac{1}{2} = p(-a, b)$$

where "$-a$" (that is, "non-a") stands for the event "reflection". Now let the experiment be carried out with one single photon. Then the probability wave packet attached to this photon will split, and we shall have the *two* wave packets, $p(a, b)$ and $p(-a, b)$, for which our equation

$$p(a, b) = \tfrac{1}{2} = p(-a, b)$$

will hold. "After a sufficient time", HEISENBERG writes, "the two parts will be separated by any distance desired ...". Now let us assume that we "find", with the help of a photographic plate, that the photon (which is indivisible) was reflected. (HEISENBERG says that it is "in the reflected part of the packet", which is a misleading metaphor.) "Then the probability", he writes, "of finding the photon in the other part of the packet *immediately becomes zero.* The experiment at the position of the reflected packet thus exerts a kind of action (reduction of the wave packet) at the distant point occupied by the transmitted packet, and one sees that this action is propagated with a velocity greater than that of light." (Cp. [26], p. 39; the italics are mine.)

Now this is the great quantum muddle with a vengeance. What has happened? We had, and still have, the relative probabilities

$$p(a, b) = \tfrac{1}{2} = p(-a, b).$$

If we take the information $-a$ (which says that the particle has been reflected), then relative to this information we get

$$p(a, -a) = 0, \qquad p(-a, -a) = 1.$$

The first of these probabilities or wave packets is indeed zero. But it is quite wrong to suggest that it is a kind of changed form of the original packet $p(a, b)$ which *"immediately becomes zero".* The original packet $p(a, b)$ remains equal to $\frac{1}{2}$, which is to be interpreted as meaning that if we repeat our original experiment, the virtual frequency of photons being transmitted will equal $\frac{1}{2}$.

And $p(a, -a)$, which is zero, is quite another relative probability: it refers to *an entirely different experiment* which, although it begins like the first, ends *according to its specification* only when we find (with the help of the photographic plate) that the photon has been reflected.

No *action* is exerted upon the wave packet $p(a, b)$, neither an action at a distance nor any other action. For $p(a, b)$ is the propensity of the state of the photon relative to the original experimental conditions. This has not changed, and it can be tested by repeating the original experiment.

It might be thought that it is unnecessary to repeat all this after 32 years. But more recently, HEISENBERG has suggested that the reduction of the wave packet is somewhat similar to a quantum jump. For, on the one hand, he speaks of "the reduction of wave-packets" as "the fact that the wave function ... changes discontinuously", adding, "It is well known that the reduction of wave-packets always appears in the Copenhagen theory when *the transition is completed from the possible to the actual* ..." that is, when "the actual is selected from the possible, which is done by the '*observer*' ...". On the other hand, he speaks on the next page of the "element of discontinuity [in] the world, which is found everywhere in atomic physics ... [and which in] the usual interpretation of quantum theory ... *is contained in the transition from the possible to the actual*". (Cp. [*27*], pp. 23 f.; the italics are mine.)

Yet the reduction of the wave packet clearly has nothing to do with quantum theory: it is a trivial feature of probability theory that, whatever $p(a, b)$ may be, $p(a, a) = 1$ and (in general) $p(-a, a) = 0$.

Assume that we have tossed a penny. (The example is taken from p. 69 of my [*55*].) The probability of each of its possible states equals $\frac{1}{2}$. As long as we don't look at the result of our toss, we *can* still say that the probability will be $\frac{1}{2}$. If we bend down and look, it suddenly "changes": one probability becomes 1, the other 0. Was there a quantum jump, owing to our looking? Was the penny influenced by our observation? Obviously not. (The penny is a "classical" particle.) Not even the probability (or propensity) was influenced. There is no more involved here, or in *any* reduction of the wave packet, than the trivial principle: if our information contains the result of an experiment, then the probability of this result, relative to this information (regarded as part of the experiment's specification), will always trivially be $p(a, a) = 1$.

This explains also what is valid in VON NEUMANN's principle, mentioned in my seventh thesis above, that if we repeat a measurement at once, then the result will be the same with certainty. Indeed, it is quite trite that if we look at our penny a second time, it will still lie as before.

And more generally: if we take a "measurement" like that of the arrived photon as *defining the conditions of the experiment*, then the outcome of the repetition of this experiment is certain, by virtue of the specification of the experiment together with the trite fact that $p(a, a) = 1$.

Before proceeding to my next thesis, I will just return for a moment to the pin board. Take

$$p(q_2, e_1) = r$$

to be the probability of a ball hitting the pin q_2, in the original experiment, and assume that we see the ball passing q_2 *without* hitting it. Then this can be interpreted exactly as HEISENBERG interprets the experiment with the semi-transparent mirror: we could say (it would be very misleading) that the wave packet $p(q_2, e_1)$ collapses, that it becomes zero with super-luminal velocity. I hope that the absurdity of the muddle need not be further elaborated.

10. My tenth thesis is that the propensity interpretation solves the problem of the relationship between particles and their statistics, and thereby that of the relationship between particles and waves.

DIRAC writes: "Some time before the discovery of quantum mechanics people [EINSTEIN, VON LAUE] realized that the connection between light waves and photons must be of a statistical character. What they did not clearly realize, however, was that the wave function gives information about the probability of *one* photon being in a particular place and not the probable number of photons in that place." (Cp. [*14*], p. 9.) And he continues with an example very much like the example discussed above of the semi-transparent mirror interacting with *one* photon.

Now this application of probability theory to single cases is precisely what the propensity interpretation achieves. But it does not achieve it by speaking about particles or photons. *Propensities are properties of neither particles nor photons nor electrons nor pennies.* They are *properties of the repeatable experimental arrangement* — physical and concrete, in so far as they may be statistically tested (and may lead, in the pin board case, to an actual characteristic physical arrangement of balls) — and abstract in so far as *any particular experimental arrangement* may be regarded as an instance of more than one specification for "its" repetition. (Take the tossing of a penny: it may have been thrown 9 feet up. Shall we say or shall we not say that this experiment is repeated if the penny is thrown to a height of 10 feet?) It is this *relativity* of the propensities that makes them sometimes look "unreal": it is the fact that they refer both to single cases *and* to their virtual repetitions, and that any single case has so many properties that we cannot say, just by inspection, which of them are to be included among the specifications defining what should be taken as "our" experiment, and as "its" repetition.

But this is true not only of all propensities or probabilities (classical or quantum-mechanical); it is also true of *all physical or biological experiments*, and it is one of the reasons why experimentation is impossible without theory: what seems to be completely irrelevant in one experiment e_1, or arbitrarily variable in its repetitions, may turn out in "another" experiment e_2 (otherwise indistinguishable) to be part of its most important specifications. Every experimentalist can give countless examples. Some so-called "chance discoveries" have been made by getting unwanted, or unexpected, results upon repeating an experiment, and then noticing that the change in the result depended upon some factor previously conjectured to be irrelevant, and therefore not included in (nor excluded by) the specification of the experiment.

Thus the *relativity to specification* of which we have spoken is characteristic neither of quantum experiments nor even of statistical experiments: *it is a permanent feature of all experimentation*. (And a propensity relation might be regarded, and intuitively understood, as a generalization of a "causal" relation, however we may interpret "causality".) For this reason it seems to me mistaken to regard statistical laws, statistical distributions, and other statistical entities, as non-physical or unreal. Probability fields are physical, even though they depend on, or are relative to, specified experimental conditions. (Cp. [49], pp. 213 f.)

11. My eleventh thesis is this: even though both the particles and the probability fields are real, it is misleading (as LANDÉ rightly insists) to speak of a "duality" between them: the particles are important *objects* of the experimentation; the probability fields are propensity fields, and as such important *properties of the experimental arrangement*, and of its specified conditions.

A simple example (taken from p. 89 of my [55]) may illustrate this. One is easily tempted to look upon the probability $\frac{1}{2}$ as a propensity of a homogeneous coin with a head and a tail side — that is, as a property of a thing, of a kind of "particle". *But the temptation must be resisted.* For let us assume an experimental arrangement in which the penny is not spun but tossed in such a way that it falls on a table with some slots in which it can be caught upright. Then we may distinguish between *three* possibilities: heads showing up; tails showing up; and neither showing up. Or even four possibilities: if the slots are all north-south, we may distinguish the caught pennies by the direction in which their heads face (east or west). This shows that conditions other than the structure of the penny (or the particle) may greatly contribute to the probability or propensity: the whole experimental arrangement determines the "sample space" and the probability distribution. (We also can easily conceive of specifications according to which the experimental conditions change, perhaps even in a certain "random" manner, while the experiment proceeds.)

Thus the propensity or probability is not (like baldness, or charge) a property of the member of the population (man, particle) but somewhat more like the popularity (and consequently, the sales statistic) of a certain brand of chocolate, depending on all kinds of conditions (advertisement, sales organization, statistical distribution in the population of preferential taste for various kinds of chocolate). And a wave-like distribution of a probability (or a probability amplitude) is, indeed, something which cannot be said to be an alternative "picture" of the member of the population (man; bar of chocolate; particle). It would be awkward to speak of a "duality" (a *symmetrical* relation) between a bar of chocolate and the shape of the distribution curve of its propensity to be sold tomorrow.

12. My twelfth thesis is that the mistaken idea of a duality of particle and wave is, partly, due to the hopes raised by DE BROGLIE and SCHRÖ-DINGER of giving a wave theory of the *structure of particles.*

There was a span of over two years between the beginning of wave mechanics and the successful analysis and interpretation of experiments as tests of BORN's statistical interpretation, first presented in 1926, of the ψ-function. (Cp. [*10*], p. 104.) In these years, the statistical problems seemed less important than the hope of solving the problems of atomic stability (and of quantum jumps) by a classical method — a very beautiful method, and an inspiring hope: the hope was nothing less than one of explaining matter and its structure by field concepts. When later SCHRÖDINGER and ECKART showed the (far-reaching though not complete) equivalence of the wave theory and HEISENBERG's particle theory, the two-picture interpretation was born, with its idea of a symmetry or duality between particle and wave. But so far as there was an equivalence, it was one between *two statistical theories* — a statistical theory ("matrix mechanics") which started from the statistical behaviour of particles, and a statistical theory which started from the wave-like shape of certain probability amplitudes. We might say (being wise after the event) that SCHRÖDINGER's hope that what he had found was a wave theory of the structure of matter should not have survived (cp. [*58*]) the successful tests of BORN's statistical interpretation of the wave theory.

13. My thirteenth and last thesis is this. Both classical physics and quantum physics are indeterministic. (Cp. [*52, 40*], and [*10*], pp. 107 to 110.) The peculiarity of quantum mechanics is the principle of the super-position of wave amplitudes — a kind of *probabilistic dependence* (called by LANDÉ "*interdependence*") that has apparently no parallel in classical probability theory. To my way of thinking, this seems to be a point in

favour of saying that propensities are physical and real (though *virtual*, as stressed by FEYNMAN). For the superposition can be kicked: coherence (the phase) can be destroyed by the experimental arrangement.

ALFRED LANDÉ has made a most interesting and it seems at least partly successful attempt to explain this peculiarity by showing mathematically that "The question ... *why do the probabilities interfere?* can ... be answered: they have no other choice if they 'want' to obey a *general interdependence law* at all." (Cp. [*38*], p. 82; the italics are partly mine.) Let us assume that LANDÉ's brilliant derivations of quantum theory from non-quantal principles of symmetry stand up to critical analysis: even then it seems to me that his own arguments show that these probabilities (propensities) whose amplitudes can interfere should be conjectured to be physical and real, and not merely a mathematical device (as he sometimes seems to suggest). Though their mathematical "pictures" may have the shape of "waves" only in "configuration space", *as propensities* they are physical and real, quite independently of the question whether or not they can be represented by a wave picture, or a function with a wave shape, or, indeed, by any picture or shape at all. The wave picture may thus have only a mathematical significance; but this is not true of the laws of superposition which express a real probabilistic dependence.

On the other hand, it seems to me clear from the Compton-Simon photographs that *photons can be kicked and can kick back*, and are therefore (in spite of LANDÉ's sceptical views as to their existence) "*real*" in precisely the sense which LANDÉ himself has given to the term.

As always, nothing depends on words, but talking of "dualism of particle and wave" has created much confusion, as LANDÉ rightly emphasizes; so much so that I wish to support his suggestion to abandon the term "dualism". I propose that we speak instead (as did EINSTEIN) of the particle and its "*associated*" propensity fields (the plural indicates that the fields depend not only on the particle but also on other conditions), thus avoiding the suggestion of a symmetrical relation.

Without establishing some such terminology as this ("*association*" in place of "*dualism*") the term "dualism" is bound to survive, with all the misconceptions connected with it; for it does point to something important: the association that exists between particles and fields of propensities ("forces", decay propensities, propensities for pair production, and others).

Incidentally, among the misleading fashionable terms of the theory is the term "observable". (Cp. [*2a*], especially pp. 465 f.) It suggests something that does not exist: all "observables" are *calculated and inferred on theoretical grounds,* rather than observed or directly measured. Thus what is "observable" always depends upon the theory we use. However, here again one should not quarrel about words; no more about the word "observable" than about the word "real". Definitions, as usual, lead nowhere; but most of us know what we mean when we say that *there are* such things (observable things) as elephants, or electrons, or magnetic fields; or (more difficult to observe) propensities, such as the propensity to attract, or to understand, or to criticize; or the propensity of an experiment to yield some specified result.

14. To sum up. The alleged dualism of particle and wave and the subjective interpretation of probability, with which it is closely connected, are responsible for the subjectivistic and anti-realistic interpretation of quantum theory and for such characteristic statements as Wigner's, who says that "the laws of quantum mechanics itself cannot be formulated ... without recourse to the concept of consciousness" (cp. [61], p. 232); a view that he attributes also to von Neumann; or Heisenberg's statement: "The conception of objective reality ... has thus evaporated ... into the transparent clarity of a mathematics that represents no longer the behaviour of particles but rather our knowledge of this behaviour." (Cp. [28], p. 100.) Or his assertion that if the observer is exorcized, and physics made objective, the ψ-function "contains no physics at all". (Cp. [27], p. 26.)

I have often argued in favour of the evolutionary significance of consciousness, and its supreme biological role in grasping and criticizing ideas. But its intrusion into the probabilistic theory of quantum mechanics seems to me based on bad philosophy and on a few very simple mistakes. These, I hope, will soon be forgotten, while the great physicists who happened to commit them will be for ever remembered by their marvellous contributions to physics: contributions of a significance and depth to which no philosopher can aspire.

References

[1] Bohm, D.: Quantum theory. 1951.

[2] — A suggested interpretation of quantum theory in terms of "hidden" variables. Phys. Rev. **85**, 166—179, 180—193 (1952).

[2a] —, and J. Bub: A proposed solution of the measurement problem in quantum mechanics by a hidden variable theory. Rev. Mod. Phys. **38**, 453—469 (1966).

[3] Bohr, N.: Can quantum-mechanical description of physical reality be considered complete? Phys. Rev. **48**, 696—702 (1935).

[4] — Discussion with Einstein on epistemological problems in atomic physics, [57], 201—241.

[5] — On the notions of causality and complementarity. Dialectica **2**, 312—319 (1948).

[6] — The quantum postulate and the recent development of atomic theory. Nature **121**, 580—590 (1928).

[7] —, u. L. Rosenfeld: Zur Frage der Meßbarkeit der elektromagnetischen Feldgrößen. Det Kgl. Danske Videnskabernes Selskab, Mathematisk-fysiske Meddelelser **12**, No. 8 (1933).

[8] BOPP, F.: Statistische Mechanik bei Störung des Zustands eines physikalischen Systems durch die Beobachtung, [9], 128—149.

[9] — (ed.): Werner Heisenberg und die Physik unserer Zeit. 1961.

[10] BORN, M.: Bemerkungen zur statistischen Deutung der Quantenmechanik, [9], 103—118.

[11] — Scientific papers presented to Max Born. 1953.

[12] BUNGE, M.: Strife about complementarity. Brit. J. Phil. Sci. 6, 1—12 and 141—154 (1955).

[13] — (ed.): The critical approach to science and philosophy. 1964.

[14] DIRAC, P. A. M.: The principles of quantum mechanics, 4th ed. 1958.

[15] ECKART, C.: Operator calculus and the solution of the equations of quantum dynamics. Phys. Rev. 28, 711—726 (1926).

[16] EDDINGTON, A.: Relativity theory of protons and electrons. 1936.

[17] EINSTEIN, A.: Sidelights on relativity. 1922.

[18] — Quanten-Mechanik und Wirklichkeit. Dialectica 2, 320—324 (1948).

[19] — Remarks concerning the essays brought together in this co-operative volume, [57], 665—688.

[20] — Zur Allgemeinen Relativitätstheorie. Sitzungsberichte der Preußischen Akademie der Wissenschaften 1915, Teil 2, 778—786 und 799—801.

[21] —, B. PODOLSKY, and N. ROSEN: Can quantum-mechanical description of physical reality be considered complete? Phys. Rev. 47, 777—780 (1935).

[22] FEIGL, H., and G. MAXWELL (eds.): Current issues in the philosophy of science. 1961.

[23] FRISCH, O. R.: Observation and the quantum, [13], 309—315.

[24] HAMBLIN, C. L.: The modal "probably". Mind 68, 234—240 (1959).

[25] HANSON, N. R.: Are wave mechanics and matrix mechanics equivalent theories?, [22], 401—425.

[26] HEISENBERG, W.: The physical principles of the quantum theory. 1930.

[27] — The development of the interpretation of the quantum theory, [47], 12—29.

[28] — The representation of nature in contemporary physics. Daedalus 87, 95—108 (1958).

[29] — Über quantentheoretische Kinematik und Mechanik. Mathematische Annalen 95, 683—705 (1926).

[30] HILL, E. L.: Comments on Hanson's "Are wave mechanics and matrix mechanics equivalent theories?", [22], 425—428.

[31] JÁNOSSY, L., and L. NAGY: Experiments on the Rossi curve. Acta Phys. Acad. Sci. Hung. 6, 467—484 (1957).

[32] —, T. SÁNDOR, and A. SOMOGYI: On the photon component of extensive air showers. Acta Phys. Acad. Sci. Hung. 6, 455—465 (1957).

[33] KNEALE, W. C.: Probability and induction. 1949.

[34] KÖRNER, S. (ed.): Observation and interpretation in the philosophy of physics, 1957. Dover ed. 1962.

[35] LANCZOS, C.: Albert Einstein and the cosmic world order. 1965.

[36] LANDÉ, A.: Foundations of quantum theory. 1955.

[37] —, From dualism to unity in quantum mechanics. 1960.

[38] —, New foundations of quantum mechanics. 1965.

[39] —, Quantum mechanics. 1951.

[40] —, Probability in classical and quantum theory, [11], 59—64.

[41] MACH, E.: The science of mechanics, 5th ed., 1941 (1st German ed. 1883).

[42] MEHLBERG, H.: Comments on Landé's "From duality to unity in quantum mechanics", [22], 360—370.

[43] MILLIKAN, R. A.: Electrons (+ and −), protons, photons, neutrons and cosmic rays. 1935.

[44] — Time, matter, and values. 1932.

[45] MISES, R. VON: Probability, statistics, and truth, 1939 (1st German ed. 1928; 3rd, 1951).

[46] NEUMANN, J. VON: Mathematical foundations of quantum mechanics, 1949, 1955 (German ed. 1932).

[47] PAULI, W. (ed.): Niels Bohr and the development of physics. 1955.

[48] POLANYI, M.: The logic of personal knowledge. 1961.

[49] POPPER, K. R.: Conjectures and refutations. 1963 and 1965.

[50] — The logic of scientific discovery. 1959 (1st German ed. 1934; 2nd 1966).

[51] — Creative and non-creative definitions in the calculus of probability. Synthese 15, 167—186 (1963).

[52] — Indeterminism in quantum physics and in classical physics. Brit. J. Phil. Sci. 1, 117—133 and 173—195 (1950).

[53] — Probability magic, or knowledge out of ignorance. Dialectica 11, 354—374 (1957).

[54] — The propensity interpretation of probability. Brit. J. Phil. Sci. 10, 25—42 (1959).

[55] — The propensity interpretation of the calculus of probability, and the quantum theory, [34], 65—70 and 88f.

[56] ROSENFELD, L.: Misunderstandings about the foundations of quantum theory, [34], 41—45.

[57] SCHILPP, P. A. (ed.): Albert Einstein: Philosopher-scientist. The library of living philosophers. 1949.

[58] SCHRÖDINGER, E.: The general theory of relativity and wave mechanics, [11], 65—74.

[59] — Über das Verhältnis der Heisenberg-Born-Jordanschen Quantenmechanik zu der Meinen. Ann. Physik 79, 734—756 (1926).

[60] SHACKLE, G. L. S.: Decision, order, and time in human affairs. 1961.

[61] WIGNER, E. P.: The probability of the existence of a self-reproducing unit, [48], 231—238.

Acknowledgment. I wish to thank my Research Assistant DAVID MILLER for his indefatigable help given so freely in connection with this paper.

Chapter 2

The Problem of Physical Reality in Contemporary Science

Henry Mehlberg *

Department of Philosophy, The University of Chicago, Ill., USA

1. The New Status of Physical Reality at the Quantum Level

In this paper I shall explore some philosophical aspects of the present scientific view of the physical reality of the micro-cosmos. Everybody remembers, of course, that some of the most insightful and least speculative thinkers of this century have stressed the physical reality of the micro-cosmos. i. e., the physical sub-universe which is presently conceived as consisting of molecules, atoms and elementary particles. H. POINCARÉ[1] argued that molecules are real since the number of molecules in a particular region can often be reliably determined: he felt that it would be impossible to count non-existent or unreal things, In 1927 P. W. BRIDGMAN [1] wrote about the status of the atom: "we are now as convinced of its physical reality as of our hands and feet". The men who keep devising increasingly powerful atomic weapons, including those who have already used them twice, probably feel in the same way.

At present, the question of physical reality arises neither at the level of molecules, which were of concern to POINCARÉ, nor at the level of atoms discussed by BRIDGMAN. We are now primarily interested in the status of physical reality at the level of elementary particles, e.g., electrons, positrons or photons. A crucial theory dealing with these three types of elementary particles and their interactions with electromagnetic fields has now come of age, mainly due to the contributions of DYSON, FEYNMAN, SCHWINGER and TOMONAGA [2]. Of course, I have in mind contemporary quantum electrodynamics which satisfies the covariance requirement, gives a reasonable amount of evidence in support of its internal consistency and provides an experimental "fit" of an unsurpassed

---

* The research summarized in this paper was done under a grant of the National Science Foundation.

[1] Cf. A. EINSTEIN: Zur Theorie der Brownischen Bewegung. Annalen der Physik, Ser. H, Vol. **19**, 371—381 (1906).

accuracy. The physical reality of electrons, positrons and photons, including the reality of the momenta, energies and polarizations ascribable to them, can be supported by an argument exactly duplicating the Poincaré argument in favor of molecular reality. One can argue that electrons, positrons and photons possessed of specifiable momenta, energies and polarizations must exist since they can be counted. The possibility of determining the numbers of any of these particles assumed to possess any specifiable momentum, energy and polarization properties is made apparent by the well-established Feynman rules [3] concerning the matrix-elements (in other words, the amplitudes of transition probabilities) for any process involving the three types of micro-systems and their associated fields. According to Feynman, any matrix-element of this type corresponding to a particular system σ of elementary particles with their respective fields involves several, physically significant integers: (1) The number of electron-lines external with regard to σ which imply a specified momentum and polarization. (2) The number of external photon-lines with a specified energy, polarization, and, possibly, with a specified external electromagnetic field. Further (3) the numbers of internal lines of the three types of particles involved in every matrix-element, in addition to (4) the number of closed electron-loops with an even number of electron-lines.

There is no point, obviously, in questioning the cogency, in quantum electrodynamics, of the sort of argument which Poincaré used. Consequently, it may seem that the physical reality of the elementary particles and of their associated fields should be granted an experimental reality status equal to the status of molecules as evaluated by Poincaré, or the reality-status of atoms according to Bridgman. Yet, a crucial and incessantly growing area of contemporary science which includes the whole of quantum physics, quantum chemistry, and quantum biology is usually interpreted by the world's scientific community as incompatible with the above view of the reality of the microcosmos. According to this conventional interpretation of all quantal theories, which is traceable mainly to N. Bohr [4] and Heisenberg [5] (and, therefore, often referred to as the "Copenhagen Interpretation"), the concept of an observer-independent physical reality is inapplicable in principle and foreign to the microcosmos [6]. Of course, the surrender of physical reality at the quantum level is not the only departure recommended in Copenhagen from the philosophical outlook which dominated the growth of science from Newton to Einstein. The abandonment of strict determinism at the quantum level [7] is another venture on which science has embarked since the Copenhagen Interpretation came into the picture in 1927, the very year when Bridgman made his statement about the physical reality of the atom.

The fact that strict determinism did not survive the quantum jump from pre-quantum science to quantum science has been discussed over and over again. However, the new scientific status of physical reality has barely been noticed by the makers of the Copenhagen Interpretation and by men like EINSTEIN, DIRAC and PAULI who contributed as much as BOHR and HEISENBERG to the emergence and growth of quantal theories. In general, however, philosophers and physicists interested in the philosophy underlying all quantum theories have apparently disregarded the physicist's switch to his present stand on physical reality, although this change of mind is philosophically more intriguing than the abjuration of determinism.

The impossibility of using the classical concept of physical reality was variously formulated by BOHR, DIRAC, PAULI, MARGENAU and other investigators. DIRAC pointed out that, at the quantum level, only exceptional circumstances enable us to assert meaningfully that a physical system *has* a given property, in contrast to the most frequent situations where only a probability of *manifesting* this property can be attributed to the system [8]. MARGENAU's distinction between *"possessed"* properties of classical systems and *"latent"* properties of quantal systems comes pretty close to DIRAC's dichotomy [9]. PAULI (and, more frequently, VON WEIZSÄCKER [10]) contrasted the *"unobjectifiability"* of quantal properties with the *"objectifiability"* of corresponding properties of the macro-level. BOHR[2] himself stated more informally, but quite intelligibly, the new stand on physical reality. I shall start with quoting and clarifying his formulation:

"We must renounce (as BOHR put it) ascribing conventional physical attributes to micro-objects in an absolute way". Three concepts in BOHR's statement call for a comment: The concept of a *micro-object*, of a *conventional* physical attribute, and of ascribing an attribute in an *absolute* way. A micro-object meant to BOHR what it still means to us: anything that is either a molecule, or an atom or an elementary particle. A conventional physical attribute in BOHR's parlance, is synonymous with what is technically called a dynamical variable. Such attributes can be exemplified by the spatial position of a physical system, its linear or angular momentum, its energy etc. Finally, on closer analysis, the recommendation to refrain from ascribing these attributes to micro-objects in an absolute way comes to prohibiting the extension to the quantum level of the classical view that the particular position or momentum of a physical system is not conditional on the actual outcome of a relevant measurement performed on the system or an actual observation of the system.

BOHR's ban on this extension to the microcosmos of the classical way of handling physical attributes epitomizes the philosophy of the Copenhagen Interpretation. On closer examination, the ban may be construed as either an epistemological or ontological principle, i. e. as dealing either with what can be *known* about the microcosmos or with what there *is* in the microcosmos. Let us start our analysis with the ontological version of the Copenhagen Interpretation. It may be paraphrased as the Principle of the Unreality of Unobserved Micro-Events. In the context of this paraphrase, a "micro-event E" is the fact that a micro-system σ has the physical property π at a particular time t. The micro-event E is said to be "unobserved" if the occurence of E is not the consequence of the outcome of a measurement performed on σ or an observation of σ [11—13]. The ontological principle of the unreality of unobserved micro-events comes to asserting that there are no unobserved events of the type E [14, 15].

To reassure those readers who prefer to disregard the ontology inherent in any theory [16] or are allergic to terms like "unreality", it may be worth mentioning that the ontological unreality principle of the Copenhagen Interpretation can be replaced with an equivalent epistemological principle. To formulate the latter, we have to recall some conclusions of earlier investigations into the meaning of knowledge, truth, extra-linguistic reference, and empirical verifiability [17]. Thus "X knows that the statement S about object σ is true" can be shown to mean that both S is true and that X was induced to believe in S by the results of his observations and measurements of σ, provided that these results constitute jointly a body of evidence adequately supporting S. The meaning of asserting that S is about (refers to) the micro-object σ or that S ascribes the property π to σ can be explained along the same lines. The assertion that "X empirically verifies the statement S about the object σ" means that the observations and measurements made by X on σ and used by him as premises in a correct logical argument have given him a knowledge of the truth-value of S. Finally, the empirical verifiability of a statement S means that the empirical verification of S by any person X would not violate any laws of nature.

In terms of these concepts, the epistemology underlying the Copenhagen Interpretation can be expressed as follows: "No statement S ascribing to a micro-object σ the property π is true unless an empirical verification of S was actually carried out". This assertion may be called the "Principle of Truth-Indeterminacy of Empirically Unverified Statements about Micro-objects". This assertion epitomizes the Copenhagen Interpretation and clearly contradicts the epistemology of pre-quantal science. There were two schools of thought in this classical epistemology: the positivistic and the anti-positivistic one. The positivists granted a specific truth-value only to empirically verifiable statements. In many cases, however, scientists opposed the positivistic restriction on truth and granted a definite truth-value also to empirically unverifiable statements. However, regardless of a scientist's stand on the issue of positivism during the pre-quantal history of science, nobody seems to have then advocated the correctness of the ultra-positivistic epistemology underlying the Copenhagen Interpretation. According to the Copenhagen school of thought, the truth-determinacy of a statement is conditional on the actual empirical verification of the statement rather than on its verifiability.

Both the new epistemology of the Copenhagen Interpretation and its ontological equivalent clash with the perennial scientific outlook and with most ideas which even the protagonists of the Copenhagen Interpretation could not help to entertain since they have been thinking within the conceptual framework inherent in this outlook. Thus, the important hypothesis that the nucleus of every atom (with the exception of hydrogen atoms) consists of protons and neutrons is creditable to HEISENBERG himself. This hypothesis ascribes to the micro-systems classifiable as neutrons or protons the conventional physical attribute of occupying a spatial position inside a nucleus. Obviously no measurement or observation supporting directly this positional attribute of either kind of nucleon was ever made. If literally interpreted, HEISENBERG's nuclear hypothesis clearly contradicts the Unreality Principle. The only way out of the difficulty would be a nonliteral reinterpretation of the hypothesis. I am not aware, however, of any attempt at providing this vital hypothesis with some acceptable metaphorical meaning. This holds also of a set of other fundamental hypotheses associated with all quantum theories.

In view of the serious difficulties raised by the Unreality Principle, it is important to determine the reasons for including it in the Copenhagen Interpretation. There are many quantum laws (e. g. the law of space-quantization to be discussed shortly) which *suggest* the Unreality Principle. It seems to me, however, that no attempt was ever made at proving the Principle from quantum mechanical axioms. In VON NEUMANN's treatise, which includes a pioneering quantum theory of measurement or observation, considered usually as an integral part of the Copenhagen Interpretation, the Unreality Principle or any equivalent formulation of it is never mentioned. However, it seems to me that the Unreality Principle can be shown to follow from VON NEUMANN's axioms if use is made of a theorem [*18*] to which I shall refer in the sequel as the "Strong Indeterminacy Principle". The theorem was proved a few years after the discovery of HEISENBERG's classical and less comprehensive Indeterminacy Principle.

The theorem can be stated as follows: Let a, b, c, d, be four real numbers satisfying the inequalities $b > a$ and $d > c$. Let $P(a, b)$ be the property possessed by any system σ such that σ is within a sphere of radius b but outside a concentric sphere of radius a. Similarly, let

$M(c, d)$ denote the property possessed by every system σ such that the modulus of the linear momentum of σ exceeds c and is smaller than d. Under these assumptions, the Strong Indeterminacy Principle asserts that the Hermitean operators on the Hilbert space associated with σ and corresponding to the properties $P(a, b)$ and $M(c, d)$, respectively, do not commute with each other for any quadruple of numbers satisfying the above inequalities. According to the standard interpretation of the quantum theory, no system σ can have at the same time both properties $P(a, b)$ and $M(c, d)$, regardless of how the numbers b and d are chosen. This means, however, that no matter how unsharply a position and a corresponding momentum be defined, no system can have either unless the outcome of a measurement of either $P(a, b)$ or $M(c, d)$ performed on σ entails, in conjunction with other data and quantum theoretical laws, that σ has the property measured. If none of these two properties were measured on σ then σ could not have any P or M-property whatsoever. For having some property P would prevent σ from having any property M and vice versa. In other words, being in any region, however vast, would prevent σ from either being at rest or moving at any conceivable speed. An electron known to be in some state of motion could not possibly be in any spatial region (if accelerated in Brookhaven, it could not possibly be in Brookhaven).

A correlation of measurements of energy and time is usually held to resemble the correlation of the position and momentum measurements. Consequently, no microevent could happen at time t unless an actual outcome of a measurement entailed such a consequence. In conjunction, the mensural correlations of the two pairs of quantities prove that nothing unobserved happens at any place and at any time. This, however, is the principle of the unreality of unobserved micro-events.

It should be pointed out that, contrary to frequently held views, the Strong Indeterminacy Principle and the ensuing limitations of the concept of reality hold for any statement S ascribing a property π to a quantal system σ regardless of whether π is attributed to σ with certainty or with a probability $p < 1$. To illustrate the probabilistic case, let us recall the law of space-quantization [19]. Thus, when a silver atom in an appropriate quantum state is placed in a horizontal, non-homogeneous magnetic field, the law of space-quantization has two consequences. (1) If we check on whether the atom is horizontally oriented, we shall certainly obtain a positive result. (2) If we check, instead, whether the atom is vertically oriented, we can expect a positive result with a probability $p > \frac{1}{2}$. Hence, if we refrain from checking on the atom's spatial orientation, it cannot have either orientation, because horizontal and vertical orientations are incompatible with each other within and without quantum theory.

To illustrate the relevance of the Unreality Principle to basic epistemological issues, we may reformulate the above example of space-quantization in human terms. Suppose that I have a friend who is single, lives in a two room apartment and does not mind if I call on the phone and then, sometimes, drop in on him. In this apartment, there is no door leading directly from one room into the other. Consequently, I have to use the door leading from the corridoor into any room if I want to enter in. I am proud to be his only friend and to have the exclusive privilege of calling him on the phone or dropping in on him. Over the years of our friendship, I found out that his indoor behavior is governed by the following statistical laws.

(1) Whenever I called him on the phone, noticed that his voice was cheerful and dropped in on him, I always found him in room 1 provided I entered this room from the corridor.

(2) However, if I first had a cheerful chat on the phone with him and then preferred to look for him in room 2, I found him in this second room in 6 out of 10 cases.

Hence, if I refrain from dropping in on my friend after a cheerful conversation on the phone, he cannot be in any of his rooms without violating two well-established statistical laws, in spite of the certainty that he does stay in his apartment. The unreality of my friend's unobserved presence in each room follows from the quantum theoretical type of laws governing his behavior.

2. The Problem of Physical Reality
in Non-Relativistic Quantum Mechanics

In this section, I shall try to modify the Copenhagen Interpretation to the extent necessary to establish the observer-independence of space-time within non-relativistic quantum mechanics. This is imperative because the space-time continuum is common to this particular quantum theory and to other, more comprehensive quantum theories discussed in section 3. The proposed modification of the Copenhagen Interpretation consists in supplementing it rather than in changing it. Let us recall that, according to BOHR's formulation of the Principle of Unreality, only conventional physical attributes of micro-systems are subject to this principle. He did not consider, however, the manifold of genuinely new, *unconventional* attributes which are used in quantum mechanics without being affected by the Unreality Predicament. These unconventional attributes include, e. g., the quantum state of a micro-system, the rate of change of this state, the transition-probability proper to a micro-system σ and equal to the incidence of processes involving the replacement of a particular quantum state of σ with another quantum state etc. We shall

4*

see that an exploration of these unconventional attributes yields the required solution of the problem of unreality within non-relativistic quantum mechanics.

To begin with, let us notice that the probabilistic laws of this theory are often interpreted as applicable only to ensembles of observed or measured values of micro-quantities [20]. However, in several important cases, these laws involve no observation or measurement of an ensemble of micro-systems and establish, therefore, an observer-independent distribution of the values of a micro-quantity in this ensemble. Thus, the quantum mechanical distribution of sizes in an ensemble of unexcited hydrogen-atoms (or, alternatively, the distribution of the distances separating the proton and the electron of every hydrogen atom in this ensemble) yields a non-vanishing probability of the occurrence of a size significantly exceeding the standard size of 10^{-8} cm. However, such giant atoms would necessarily have an energy exceeding the smallest possible energy common to all the atoms in the ensemble. This shows the impossibility of identifying the probability of a giant size in the ensemble with the relative frequency of the sub-ensemble of giant atoms within the entire ensemble [5]. Yet, the frequency-interpretation of quantum mechanical probabilities is accepted in all the major expositions of this theory [21—23] and seems necessary for establishing the observer-independence of the relative frequencies or incidences of the atomic sizes determined by the aforementioned distribution.

The difficulty of making the positive probability of giant size in the ensemble consistent with the minimal energy of the entire ensemble is usually solved by applying the Principle of the Unreality of Unobserved Events. It is stipulated that the quantum mechanical distribution specifies only the incidence of any particular *observed* size in the sub-ensemble of all atoms whose size has been actually measured. The incompatibility of giant size and minimal energy of a hydrogen atom is then removed because the measurement can be shown to increase the atom's energy to the extent required by its size. Obviously, the price for removing inconsistencies in this case is the application of the Principle of Unreality to quantal probabilities: only observed sizes of hydrogen atoms are supposed to be governed by the probabilistic laws of quantum mechanics.

Similarly, the value of the square of the modulus of the state function ψ of a system σ in a point defined by the set x_0, y_0, z_0, of space coordinates at the instant t_0 (symbolically, $|\psi(x_0, y_0, z_0, t_0)|^2$) is usually interpreted as the probability that σ could be observed or "found" at t_0 in point x_0, y_0, z_0. Identifying $|\psi(x_0, y_0, z_0, t_0)|^2$ with the probability of the presence of σ at t_0 in (x_0, y_0, z_0) regardless of whether the relevant measurements were actually made would lead again to serious inconsistencies. But, in problems involving quantal collisions, the value $|\psi(x_0, y_0, z_0, t_0)|^2$

must be interpreted as the probability that the system σ would *hit* another system σ at time t_0 in the point (x_0, y_0, z_0) without any measurement or observation of σ being involved.

The observer-independence of the frequencies of atomic collisions suggests a different interpretation of both the probability determined by $|\psi(x, y, z, t)|^2$ and the probability of any specific size being attributed to the aforementioned unexcited hydrogen atoms. On closer analysis, the collision probability for σ at t in point (x, y, z) comes to the probability that σ will make the transition from its momentary quantum state to another state characterized by the location of σ at time t in this point. The transition may occur as a result of a quantal collision or due to a position-measurement on σ. But measurement has no monopoly on these transitions. Similarly, the probability that an unexcited hydrogen atom has any specific size should be interpreted as the probability that this atom would effect a transition from the quantum state characterized by the attribute of minimal energy to a different state defined by any specific size.

The transition probabilities between states of radiating atoms in an atomic population show the observer-independence of the corresponding frequencies in this population more convincingly. These transition probabilities have always been interpreted as numerically equal to the intensities of radiations emitted by the atoms effecting any particular transition. The probability that a transition of a given kind will occur in the atomic population cannot be construed as the relative frequency of *observed* transitions of the kind considered. In order to effect the transition, the atoms need not be observed at all. As a matter of fact, they never were, in the experiments designed to check on the empirical validity of probabilistic quantum laws. The observer-independence of the relative frequencies of transitions effected by micro-systems follows obviously from the observer-independent status of the corresponding intensities. Furthermore, the observer-independence of these intensities is clearly consistent with the Copenhagen Interpretation because radiation-intensities are macro-physical phenomena with regard to which the applicability of this standard interretation was never claimed. This holds, for similar reasons, of the transition probabilities associated with a particular size of an unexcited atom or a particular location of a micro-system.

To sum up: since the quantal laws governing transition probabilities do not involve the concept of measurement or observation, as shown by von Neumann's general formulation of his axiom "$E\,2$" [24], the observer-independence of the corresponding frequencies of changes of micro-systems does not raise any justifiable doubt. However, the observer-independence of these quantal transition frequencies leads to a most significant conclusion: If we grant the observer-independence of transition frequencies between quantum states, we also have to grant the status of observer-independence to the quantum states of the systems which make these transitions. It stands to reason that whenever the relative frequency of transitions between state I and II in a population of micro-systems can be reliably known without any micro-system having been actually observed, every particular micro-system has ("possesses") its state no matter whether any observation thereof was made. How can a system effect the transition from state I to state II if it never has been in any of these states?

Moreover, the differences between quantal states, quantities and properties of micro-systems turn out to be merely formal, on closer analysis. A property can be interpreted as a quantity capable of two values only. A quantum state can be construed, according to a procedure used by Segal [25], as a functional defined on the expectation values of measurable quantities. The observer-independence of the former entails that the latter are also observer-independent, and vice-versa. This logical interrelatedness of the crucial concepts of a quantal state, quantal property and quantal quantity provides for the possibility of shifting the status of observer-independence from any of these concepts known to be observer-independent to the two remaining concepts.

The main point can now be made concerning the limited possibility of establishing the physical observer-independent reality of the space-time continuum within non-relativistic quantum mechanics. The result to be briefly discussed here can be stated as follows: The observer-independence of time and space follows from the observer-independence of quantum states. More specifically, only the reality of the field-like aspect of time and space can thus be established. The status of spatiotemporal attributes of micro-systems is not within the reach of the quantum mechanical argument.

Let us first notice that the observer-independence of time within non-relativistic quantum mechanics does not raise difficulties comparable to those facing the issue of space. The reason for the less controversial onto-logical status of the concept of time is that, in the quantum mechanical formalism, the time variable functions as a real number, and not as a Hermitean operator on a Hilbert space. The proof of the Indeterminacy Principle and its corollary, the Unreality Principle with regard to space, assumes that spatial coordinates and their conjugated momenta are re-interpreted as operators. This is made apparent by the fact that the crucial premise of the proof is the non-commutativity of the space and momentum operators.

One would expect, therefore, that the Unreality Principle should not have been applied to time coordinates. This has not been the case in the most authoritative presentations of quantum mechanics because the circumstance that, in prequantal physics, time was a dynamical variable conjugated with another dynamical variable, viz. energy, has induced several authors to claim that the same type of "Uncertainty Principle", which HEISENBERG had established for canonically conjugated positional and momentum variables, is also valid for the time and energy variables in quantum mechanics. This is, however, not the case, as has been shown by TAMM and MANDEL'SHTAM [26]. These physicists were able to derive a sort of Uncertainty Principle involving time and energy but had to use to this end additional premises and to formulate the Uncertainty Principle in such a way that no observer-dependence or Unreality Principle can be derived therefrom for physical time. We may conclude, therefore, that non-relativistic quantum mechanics provides no evidence in favor of the physical unreality or the observer-dependence of time. The parallel treatment of time and space by most investigators rests on a misinterpretation of their roles in the quantum mechanical formalism.

In addition to the circumstance that time is represented in the quantum mechanical formalism by a real number, we have to consider the essential fact that both time and space are uniquely associated with the observer-independent quantum state φ of a micro-system due to the structure of the function ψ which describes φ and is therefore connected by a bi-unique relation R_1 with φ. The essential properties of ψ can be stated as follows: it is a complex-valued function of four real variables whose joint ranges coincide with the spacetime continuum. This implies that the state ψ is uniquely associated by a particular, bi-unique relation R_2 with a mapping M of the space time continuum into the set of all complex numbers (or equivalently into the set of all ordered pairs of real numbers). The observer-independent existence of φ is therefore equivalent to the observer-independent existence of the mapping M. However, a mapping of any domain D into any range R cannot exist unless its domain is existent. In our case, the domain of the mapping M is the space-time continuum. Hence, the observer independence of φ entails that the space-time continuum has the same status of an observer-independent existence.

The gist of this derivation of the reality of space-time from the reality of quantum states is that the spatio-temporal coordinates function as parameters of any quantum state. All the four coordinates are c-numbers, not q-numbers in the quantum mechanical formalism. Their values are therefore unique and observer-independent.

The limitations of the reality ascribable to space-time in this theory are apparent. Space-time coordinates are unconventional attributes in BOHR's parlance since they determine a highly unconventional physical entity, viz. the quantum state φ. There is no possibility to derive, within quantum mechanics, the reality of space-time attributes related to physical systems or conceived as dynamical variables. In section 3 we shall see, however, that the reality of a field-like space-time, inherent in quantum mechanics, is also of decisive importance in all relativistic quantum field theories.

3. The Problem of Physical Reality in Relativistic Quantum Field Theories

In the preceding section, I have tried to find out to what extent the crippling epistemological and ontological restrictions on the scope of physical reality which cannot be overcome within non-relativistic quantum mechanics on its Copenhagen Interpretation may be reduced by modifying this interpretation. The impossibility of completely ridding this particular quantum theory of all its restrictions on physical reality, regardless of the interpretation put on the theory, is due to the fact that the theory is incapable of accounting for the following fundamental aspects of the physical universe: (1) The merger of space and time into a single four-dimensional continuum effected in the Special Theory of Relativity [27]. (2) The spatio-temporal pervasiveness of the basic physical quantities requiring a field-theoretical approach to the physical universe. (3) The discovery of the sub-universe of elementary particles including the unexpected variety of their masses, life-spans and their numerous quantal characteristics such as spin and strangeness, in addition to their susceptibility to creation and annihilation.

In contrast to non-relativistic quantum mechanics, several contemporary, relativistic quantum field theories are able to account, more or less satisfactorily, for these three aspects of the physical universe. Consequently, these theories are conducive to a remarkably broader concept of physical reality than non-relativistic quantum mechanics. In this discussion, it will suffice to consider one reasonably established theory of this type, viz. quantum electrodynamics. Prior to exploring the status of physical reality in quantum electrodynamics, it should be pointed out that the problem of ridding quantum electrodynamics of the epistemological and ontological limitations of the scope of physical reality is far from trivial, in spite of the substantial advances made by this theory. The reason why this problem also faces quantum electrodynamics is simple. In non-relativistic quantum mechanics, the problem arose mainly in connection with the Principle of Indeterminacy and the ensuing limitations on the measurability of quantum mechanical quantities. The theory of quantum measurement was extended to quantum electrodynamics, mainly by Bohr and Rosenfeld. They discovered limitations on the measurability of quantities occurring in this new area similar to the limitations prevailing in quantum mechanics [28—31]. Thus, several field quantities, e. g., the electrical and the magnetic field strengths, are not measurable simultaneously because they are represented, within the new quantum theory, by non-commuting Hermitean operators. This holds also of several, revelatory quantities or "observables" used extensively in quantum electrodynamics, e. g., the "particle-number" observables and various "occupation-numbers" observables. Hence, the main source of difficulties related to physical reality in quantum mechanics, reappears in the new quantum theory. Nevertheless, in the final analysis, we shall realize that the distinctive features of quantum electrodynamics, viz. its relativistic covariance, its field-like approach to physical problems and its reasonably successful use of field-quantization in the treatment of elementary particles, jointly warrant an interpretation which does not imply crippling restrictions on the physical reality of the universe as described in this theory.

In the sequel, I shall briefly discuss the relevance of the aforementioned distinctive features of quantum electrodynamics, i. e., covariance, field-like approach, and advances in the theory of elementary particles. My three consecutive comments refer to some basic properties of the logical structure of quantum field theory. It may therefore be desirable to outline this structure before making the comments on the problem of physical reality. The logical structure of quantum electrodynamics is due to the fusion of two other theories integrated in it. The fundamental axioms of quantum electrodynamics are derived by a superposition of DIRAC's relativistic quantum mechanics of electrons and positrons on MAXWELL's field theory of electrodynamics. To adjust MAXWELL's and DIRAC's sets of fundamental equations to each other, both the relativistic covariance of DIRAC's theory and MAXWELL's field-theoretical treatment of electrodynamics must be built into the foundations of quantum electrodynamics. Such a unification requires a profound change of the logical structure of MAXWELL's and DIRAC's original theories. The change of logical structure is pervasive. We shall point out some new logical aspects which are relevant to the issue of physical reality.

Thus, MAXWELL's electromagnetic potentials which can replace his two observable field-variables and proved to be more satisfactory in quantum theoretical contexts than the field observables, had been conceived in pre-quantal electrodynamics as real-valued functions falling, respectively, under the category of vectors and scalars, and defined over the entire space and time. In relativistic electrodynamics, the potentials have obviously been defined over the four-dimensional pseudo-Euclidean continuum. In quantum electrodynamics, these potentials are reinterpreted as operators over the Hilbert space associated with the compound quantal system consisting of an electromagnetic field and all the inter-acting electrons and positrons. These Hermitean operators can be specified by any set of values of the spatio-temporal coordinates. The dependence of the operators on these spatio-temporal parameters is describable by functions defined over the space-time continuum and satisfying Maxwell's partial differential equations. However, since the operators transform functions of a Hilbert space into other functions belonging in this space, they can be assumed to transform any function in the Hilbert space which describes the quantum state of the aforementioned compound quantum system into another state function. These complex-valued state functions need not be construed as mapping the spatio-temporal coordinates of every sub-system of the compound total quantum system into the set of complex numbers. It is philosophically significant that according to a procedure first used by Dirac in his semi-quantal theory of radiation, the state functions can also be construed as depending only on a complete set of eigen-values of commuting occupation-number operators associated with any particular quantum state of the entire compound quantum system. These eigen-values of the occupation-number operators may specify e. g., the numbers of all photons, electrons and protons, on the assumption that specific momenta, energies, spins etc. are ascribable to all the elementary particles involved. The Hermitean operators which represent measurable quantities and the electromagnetic potentials can be construed as either increasing or decreasing the occupation-numbers which characterize a quantum state, depending upon whether these operators are classifiable as "creation operators" or as "annihilation operators". Since the operators affect these occupation-numbers, they change also the state function.

We can now start the discussion of the relevance to the problem of physical reality of the three distinctive features of quantum electrodynamics. Let us first evaluate the import of the field-theoretical aspect of quantum electrodynamics. The significance of this field-like concept of the entire, compound quantum system is apparent from the fact that this concept determines the physical interpretation of the theory to a large extent. The interpretation of any theory which involves a mathematical formalism must obviously include the rules which assign extralinguistic physical entities to specific formulae or symbols of the formalism. In particular, these interpretive rules, often referred to in technical discussions as the "correspondence rules" of the language under consideration, must include a definition of all the individual entities which can be referred to in the language. The set of all individuals associated in this way with a language L is usually called the "universe of discourse" of L. If L is used to formulate any field theory then the universe of discourse of L consists obviously of a sufficiently comprehensive sub-set of the set of all conceivable spatio-temporal regions, including the point like regions.

The fact that the universe of discourse of the language used to formulate quantum electrodynamics coincides with the space-time continuum is highly significant. A more detailed analysis shows that the choice of a language whose universe of discourse coincides with space-time is inevitable in view of several features of quantum electrodynamics. In this context, it may suffice to point out that a satisfactory definition of the concept of an "inertial frame of reference" must be couched in a language associated with a field-like universe of discourse because this concept, indispensable both at the quantum level and the macrophysical level, becomes inherently inapplicable at the micro-level if its standard definition in terms of the macro-physical ideas of a rigid yardstick and a natural clock both at rest relative to a set of three mutually perpendicular rigid rods is applied to micro-objects. The precise definition of an inertial frame of reference, physically meaningful both at the macro- and the micro-level and expressed in a language associated with a field-like universe of discourse, can be found in a companion paper of mine [32]. The necessity of associating a field-like universe of discourse with any language we use for expressing contemporary, relativistic physical theories can also be illustrated by the fact that the complete annihilation of particles with a non-vanishing rest-mass, in conjunction with the corresponding increase of the amount of radiating energy in the universe, is compatible with the laws of nature which have so far been established. During the time-interval when only radiating energy would constitute all there is in the universe, this universe would still be governed by laws invariant under any Lorentz transformation. However, if the relativistic covariance is not interpreted as involving only an algebraic property of the equations which express the laws of nature but rather as a physically meaningful statement, then the verifiable physical changes associated by the Special Theory of Relativity with a change of an inertial frame of reference must be expressed in terms of the relations obtaining between any inertial frame of reference and the corresponding physical effects. In another words, a physically meaningful statement of the Lorentz covariance of laws of nature involves the concept of an inertial frame of reference and, consequently, the use of a language associated with a field-like universe of discourse.

The physical reasons which make the use of a language whose universe of discourse coincides with relativistic space-time inevitable in quantum electrodynamics and similar field theories provide strong evidence in support of an observer-independent reality of the spatio-temporal continuum. The scope of physical reality implied by quantum electrodynamics transcends significantly the reality of space-time which can be established within non-relativistic quantum mechanics. The advance achieved in quantum electrodynamics in establishing the status of physical reality can be characterized as follows:

(1) In quantum mechanics, the reality of space and time is warranted in view of the fact that spatial and temporal coordinates, construed as c-numbers, function as parameters in the specification of quantum states and owe to this parametric function an observer-independence comparable to the independently established observer-independence of quantum states. This parametric use of spatio-temporal coordinates is substantially extended in quantum electrodynamics. The point is that, apart from quantum states, the formalism of this field theory includes also Hermitean field operators which represent measurable quantities but involve space-time parameters construed as c-numbers and necessary for the specification of every particular operator. We may hesitate, in this case, whether the role played by space-time parameters suffices only to grant to spatio-temporal coordinates the role of "unconventional physical attributes", to use Bohr's phrase once more. Granted, these coordinates are again associated with highly unconventional entities, viz., with Hermitean operators on a Hilbert space instead of a vector in this space. However, these operators stand for physical measurable quantities, which are unconventional, in contrast to the quantum states which are associated with space-time parameters but have to be classified as highly conventional. The labels "conventional" and "unconventional" are hardly significant in a discussion of the status of physical reality. Obviously, spatio-temporal concepts are essential components of the most conventional and least controversial physical attributes within the formalism of quantum electrodynamics and this is certainly relevant to the physical reality of the spatio-temporal continuum. However, we have to admit again that only the reality of the field-like aspect of space-time can be established in this way or, alternatively, that only the reality of a relativistic, but not that of a relational space-time is implied by quantum electrodynamics [*33*]. We are going to see, however, in discussing, in the next paragraph, the effect on the issue of reality, of being committed to a language with a universe of discourse identifiable with the spatio-temporal field, that the reality of the field-aspect of space and time is actually all that matters.

(2) However, a closer analysis of the use of a language with a spatio-temporal universe of discourse which is inevitable in the formulation of quantum electrodynamics shows a considerably more significant extension of the scope of physical reality. We may call languages whose universe of discourse coincides with space-time, field-languages. In addition, a field-language L may fall under the category of an "occupation-number type of field-language" if all the predicates which are available in L and are,

obviously, applied to spatio-temporal regions, happen to be occupation-number predicates. A field language need not be also an occupation-number type of language. In quantum electrodynamics, the language used to formulate all the relevant laws and facts falls under this special category of linguistic systems.

The significance of using an occupation-number type of a field-language can be explained as follows. Since the universe of discourse of this language L coincides with space-time, every statement expressible in L is about a space-time region (if the statement is of the subject-predicate type) or about several regions, (if the statement is relational). Thus, in a language with a different universe of discourse, we attribute a particular physical property, e. g., the property of having a particular rest-mass, or a particular electric charge, or momentum, or energy to an elementary particle, or a micro-system containing several elementary particles, or to a macro-system which consists of a set of micro-objects the cardinality of which may have an order of magnitude exceeding twenty of compared to the order of magnitude of a micro-system. These statements are replaced in a field-language of the occupation-number type with physically equivalent statements which specify, for a particular space-time region, the occupation-numbers that refer to this region and are associated with the property consisting in the presence in this region of a specific number of systems with a particular rest-mass (or, preferably, a specific number of space-time points with a locally concentrated rest-mass), or a particular electric charge, etc. Thus all conceivable physical properties ascribed in a language not classifiable as a field-language to particles or systems of particles, are replaced in the occupation-number type of field-language used in quantum electrodynamics with an equivalent statement which attributes corresponding occupation-numbers to a corresponding space-time region.

Consequently, the empirical verifiability of statements made in quantum electrodynamics establishes the physical reality of the spatio-temporal continuum in a much stronger way than the parametric role of spatio-temporal coordinates could possibly warrant: spatio-temporal regions have, independently of the observer-independent role which they play as parameters specifying real physical entities, the physical reality of an entity existing in its own right. Every other physical entity can have physical reality only owing to its association with a space-time region. This does not mean that we are driven to an absolute view of space and time and to a surrender of the relativistic view. The view of space and time which is built into the very logic of quantum electrodynamics may be expressed in terms often used in philosophical discussions of space and time: the field-like language of quantum electrodynamics implies a relativistic but not a relational view of space and time. I have explained elsewhere the meaning of a relational view of the cosmic medium. In a meta-language associated with our field-language, the relational view may be summarized as the claim that, with regard to this field-language, the space-time continuum contains the only things one can talk about in the field-language and that this continuum is not conceived as a set of relations obtaining among other entities. The relativistic tenet built into this field-language comes to the claim that all the splits of the four-dimensional continuum into a time-like and space-like component which are induced by any inertial frame of reference are on a par as far as all physical theories and laws are concerned: a physical law formulated relative to any split of the continuum remains unchanged and valid relative to any other split.

Finally, let me point out that the reality of physical events each of which consists in the presence of a property π in a micro-system or a macro-system σ is also inherent in a translation of this statement made in a language whose universe of discourse consists of particles and comparable systems into a field language: the translation would assert the observer-independent presence of an occupation-number property corresponding to π in the space-time region containing σ. Thus, using a field-language does not imply the reductionist tendency to disregard the reality of events. Nor is the field language completely unrelated to ordinary, common language. The statement "It is hot here now" is of a field-linguistic variety although it is perfectly meaningful in ordinary English. We could also show that the puzzling status of our friend whose indoor behavior is governed by statistical laws of the quantal kind can be easily solved by reformulating the laws and facts related to this quantized friend in a field-language.

I shall have to refrain from commenting on the relevance of the fact that the field language of quantum electrodynamics is of the occupation-number type and discuss briefly the relevance to the problem of physical reality of the relativistic covariance and of the use of field-quantization in this theory. In discussing the implications of a field-theoretical approach, I have already pointed out that the relativistic covariance required of all laws of nature can be physically meaningful only if the algebraic property of physical laws of remaining unchanged under a Lorentz transformation is supplemented by explicitly stating the physical effects derivable from Special Relativity in terms of an inertial frame of reference. The conventional concept of an inertial reference-frame has to be redefined to be applicable at the quantum level. The only possible way of redefining it in this way leads to the choice of a field-like language. The Special Theory of Relativity therefore contributes independently to establish the physical reality of a relativistic but non-relational world-continuum because it provides another argument in support of formulating scientific theories in a field-language.

As a matter of fact, the issue of relativistic covariance has become substantially more complex in all quantal theories because the equivalence between the validity of the Lorentz group of transformations for space-time coordinates no longer guarantees the covariance of the laws of nature under a Lorentz transformation and the classical transformations of various types of physical quantities [34]. These new aspects of relativistic covariance, due to the representation of physical quantities by Hermitean operators, has been discussed in an earlier paper of mine. The new status of covariance could be shown to extend the scope of physical reality in another direction. I shall refrain from commenting on this aspect of the problem of reality and shall also skip the relevance of relativistic observables to our main issue [35]. The only additional remark which may be added in this context concerns the Principle of Micro-causality [36], which precludes the interaction of micro-events separated by a space-like interval. I have shown elsewhere that the relation obtaining between two events separated by a space-like interval provides a sufficient conceptual basis for axiomatizing relativistic space-time so as to make this space-time physically meaningful at the macro- and micro-level. This may clarify the emphasis put in my whole argument on the physical reality of space and time.

Let me wind up this discussion of the present status of physical reality with a few brief remarks on the single most significant extension of the scope of this concept which is due to the field-theoretical approach to problems concerning elementary particles. Quantum electrodynamics is obviously creditable with a remarkable part played in this extension of the scope of physical reality. In the opening section, I have already stressed the existential import of the countability of various types of elementary particles possessed of various momenta, charges, spins etc. This is a way of establishing the existence of physical entities already used by POINCARÉ in another context and applied, in section 1, to FEYNMAN'S rules concerning the matrix-elements related to various processes at the quantum level. From our present point of view, I should have rather stressed the existential import of empirically verifiable occupation-numbers. Obviously, the facts and laws concerning elementary particles contribute to the extension of observer-independent physical reality in many other ways, in addition to the fundamental role of countability in situations occurring at the quantum level.

It is a logical triviality that elementary particles could not be created or destroyed if there were no elementary particles at all. It is equally obvious that no classification of elementary particles based on their rest-mass, or charge, or spin, or strangeness or any other attribute could have been validly established if there were no set of elementary particles to classify. The growing list of discoveries concerning elementary particles and presupposing their physical reality is as impressive as any physical discovery ever was. I do not think that the crippling restrictions on the concept of physical reality which had once been derived from a brilliant interpretation devised forty years ago for a particular, obviously limited and not literally tenable quantum theory can presently be maintained and claimed for the growing list of ever more powerful quantum theories, in spite of the lip service which is still being paid to this interpretation.

References

[1] BRIDGMAN, P. W.: The logic of modern physics, p. 49. New York: MacMillan & Co. 1927.

[2] KÄLLÉN, G.: Quantenelektrodynamik. In: Handbuch der Physik, Bd. V/1. Berlin-Göttingen-Heidelberg: Springer 1958.

[3] FEYNMAN, R. P.: Theory of positrons. Phys. Rev. 76, 749 (1949).

[4] BOHR, N.: Atomic theory and the description of nature. Cambridge: Cambridge University Press 1934.

[5] HEISENBERG, W.: The physical principles of quantum mechanics. Chicago: University of Chicago Press 1930.

[6] — The Copenhagen interpretation of quantum theory. Physics and philosophy, p. 44—58. New York: Harper 1958.

[7] BUNGE, M.: Causality. The place of the causal principle in modern science. Cambridge, Mass.: Harvard University Press 1959.

[8] DIRAC, P. A. M.: The principles of quantum mechanics, p. 35 ff. Oxford: Clarendon Press 1935.

[9] MARGENAU, H.: Open vistas. Philosophical perspectives of modern science, p. 136 ff. New Haven: Yale University Press 1961.

[10] WEIZSÄCKER, C. F. v.: The world view of physics. Chicago: University of Chicago Press 1952.

[11] FEYERABEND, P. K. On the quantum-theory of measurement. In: S. KÖRNER, ed.: Observation and interpretation, p. 121 ff. New York: Academic Press 1957.

[12] DURAND, L.: On the theory of measurement in quantum mechanical systems. Phil. Sci. 27, 115—133 (1960).

[13] WIGNER, E. P.: Die Messung von quantenmechanischen Operatoren. Z. Physik 133, 101—108 (1952).

[14] LONDON, F., et E. BAUER: La théorie de l'observation en mécanique quantique. Actualités Industrielles et Scientifiques Hermann, Paris 1939.

[15] MEHLBERG, H.: The observational problem in quantum theory. Proc. XII Int. Philos. Congr. 1961.

[16] — Theoretical and empirical aspects of scientific theories. Proc. Int. Congr. Log. Meth. Philos. Science 1960.

[17] — The reach of science, p. 263 ff., 275 ff. Toronto: University of Toronto Press 1958.

[18] LUDWIG, G.: Mathematische Grundlagen der Quantenmechanik, S. 61 ff. Berlin-Göttingen-Heidelberg: Springer 1954.

[19] KOMPANEEYETS, A. S.: Theoretical physics, p. 295 ff. New York: Dower 1961.

[20] HOUTAPPEL, R. M. F., H. VAN DAM, and E. P. WIGNER: The conceptual basis and use of the geometric invariance principles. Revs. Mod. Phys. 37, 595 (1965).

[21] FEYNMAN, R. P., and A. R. HIBBS: Quantum mechanics and path integrals, p. 1 ff., 321 ff. New York: McGraw-Hill 1965.

[22] MEHLBERG, J. J.: Is a unitary approach to foundations of probability possible? Current issues in the philosophy of science 1961.

[23] POPPER, K. R.: The propensity interpretation of the calculus of probability and the quantum theory. Observation and interpretation, p. 65 ff. New York: Academic Press 1957.

[24] NEUMANN, J. v.: Die mathematischen Grundlagen der Quantenmechanik, S. 131 ff. New York: Dover 1943.

[25] SEGAL, I. E.: Postulates for general quantum mechanics. Ann. of Math. 2, 48, 930—948 (1947).

[26] MANDEL'SHTAM, L. I., i I. YE. TAMM: Izvest. Akad. Nauk. S.S.S.R. 9, 122 (1945).

[27] GRÜNBAUM, A.: Philosophical problems of space and time. New York: Alfred A. Knopf 1963.

[28] BOHR, N., u. L. ROSENFELD: Zur Frage der Meßbarkeit der elektromagnetischen Feldgrößen. Det. Kgl. dansk. Vid. Selskab. 12, 8 (1933).

[29] — — Field and charge measurement in quantum and electrodynamics. Phys. Rev. 78, 794—798 (1950).

[30] HEITLER, W.: Physical aspects of quantum-field theory in the quantum theory of fields. Int. Inst. Phys. 1961, p. 37 ff.

[31] FEYNMAN, R. P.: The present status of quantum electrodynamics, loc. cit., p. 61 ff. Int. Inst. Phys. 1961.

[32] MEHLBERG, H.: Space, time and relativity. Proc. Int. Congr. Log. Meth. Phil. Sci. Amsterdam: North.-Holland Publ. Co. 1964.

[33] — Relativity and the atom. Mind, Matter, Method., p. 449 ff. Minnesota: University of Minnesota Press 1966.

[34] GEL'FAND, I. M., P. A. MIN'OS, and Z. YA. SHAPIRO: Representations of the rotations and Lorentz groups, p. 263 ff. Oxford: Pergamon Press 1963.

[35] WIGNER, E. P.: Relativistic invariance in quantum mechanics. Nuovo. cimento 3, 517 (1963).

[36] BOGOLIUBOV, N. N., and D. V. SHIRKOV: Introduction to the theory of quantized fields, p. 200 ff. New York: Interscience 1959.

Chapter 3

The Quantum State Vector and Physical Reality[1]

Peter G. Bergmann

Department of Physics, Syracuse University, Syracuse, N. Y., U.S.A.

Ensembles of physical systems obeying quantum laws can be constructed in a variety of ways. The properties of ensembles constructed in the most common way may be summarized conveniently in terms of a single state vector in Hilbert space, or of a density matrix, and this circumstance might suggest that the quantum state represents indeed the essence of the state in which an individual physical system finds itself, just as this essence is provided in non-quantum physics by the coordinates of a point in phase space. In this paper it is shown that the properties of more general ensembles cannot be summarized in terms of state vectors or density matrices. In view of the fact that the assertions of quantum theory generally refer to statistical properties of ensembles, not to individual systems, it is contended that the identification of individual systems with specific quantum states is questionable.

This paper represents the further development of a line of exploration that was begun several years ago[2]. This earlier paper was concerned with the claim by von Neumann[3] that quantum measurements always lead to an increase of the dispersion, and hence of the entropy of ensembles of quantum systems. This monotonic increase, we showed, depends on the manner in which the ensemble is constructed. If the selection of systems to form the ensemble depends exclusively on a screening preceding in time the measurements to be performed, then the very process of measurement introduces an uncontrolled interaction of the systems with the apparatus of the experimenter, and the increase in dispersion is required by the theory, as it is intuitively acceptable. But we showed, by constructing explicitly alternative ensembles, that for one class the entropy decreases monotonically in time as the result of measurements, whereas

---

[1] This work has been supported in part by the Air Force Office of Scientific Research and by the Aerospace Research Laboratories, Office of Aerospace Research.

[2] Aharonov, Y., P. Bergmann, and J. Lebowitz: Phys. Rev. **134**, B1410 (1964).

[3] Neumann, John von: Mathematical foundations of quantum mechanics. Princeton: Princeton University Press 1955. (Translated from the German original by R. T. Beyer.)

in more general situations the change in entropy is not monotonic in either direction.

Incidental to this principal thesis we pointed out that the assignment of a quantum state to a given system appears ambiguous. The purpose of the present paper is to elaborate this argument. In the conventional quantum theory of measurement an ensemble that has been obtained by the screening of systems of a given type in accordance with the outcome of a measurement of an observable A, in which the ensemble is to consist, say, of all those systems in which A is found to have the numerical value a, one assigns to the members of this ensemble the state $|a\rangle$. (It is assumed that a is a non-degenerate eigenvalue, and similar assumptions will be made in all that follows.) That in a subsequent measurement of the observable B the eigenvalue b_k will be observed will occur in a percentage of cases given by the absolute square of the product of the two vectors $|a\rangle$ and $|b_k\rangle$,

$$P(b_k) = \langle a|b_k\rangle \langle b_k|a\rangle. \tag{1}$$

This expression is bilinear in $|b_k\rangle$ and its dual $\langle b_k|$. The coefficients of this bilinear form, the idempotent matrix $|a\rangle\langle a|$, depend only on the selection procedure employed in constructing the ensemble on which measurements are to be performed. This is why the state $|a\rangle$ appears to be characteristic for the ensemble as such, and an intrinsic property of its member systems. The replacement of the matrix $|a\rangle\langle a|$ by a more general density matrix in the event of partial incoherence is well known and does not need to concern us now. Suffice it to point out that the density matrix, also, is contingent on a screening procedure preceding in time all subsequent manipulation of the physical systems composing the ensemble, a procedure which is merely less rigorous than confinement to one non-degenerate eigenvalue of an observable.

In what follows we shall be concerned with generalizations of the screening procedures. To avoid verbal ambiguities, we shall denote all probabilistic assertions of quantum theory by the term "assertion" if no time sense is to be implied. The term "prediction" will be reserved for assertions concerning the outcome of measurements in the future, and "retrodiction" for assertions concerning past observations. The standard formula (1) refers to predictions in the narrow sense. It can be extended to apply to a series of consecutive measurements of observables $B, C, \ldots, F$. The frequency with which the eigenvalues $b_k, c_m, \ldots, f$ will be obtained is given by the expression:

$$P(b_k, c_m, \ldots, f) = \langle f| \ldots\rangle \ldots \langle c_m|b_k\rangle \langle b_k|a|\rangle \langle a|b_k\rangle \langle b_k|c_m\rangle \ldots \langle \ldots|f\rangle. \tag{2}$$

The expression (2) is still part and parcel of the conventional approach to the quantum theory of measurement in that the screening observation of A precedes in time all other determinations, about which the probabilistic assertions are being made. It incorporates, among other facets, Heisenberg's uncertainty relation, in that the probability of obtaining the specified sequence of results cannot be zero or one if adjacent observables have non-vanishing commutators. Incidentally, in order to avoid intermediate transition matrices, Eqs. (1) and (2) have been written down for observables that are constants of the motion, though not necessarily time-independent. As any observable can be converted trivially into a (time-dependent) constant of the motion, this restriction is formal rather than substantive.

Expression (2) may now be employed to obtain assertions referring to ensembles that have been obtained by means of multiple screening procedures. All the essentials are exhibited by an ensemble that consists exclusively of systems in which the initial measurement, of A, has led to the value a, and the final measurement, of F, to the value f. This ensemble is clearly a subensemble of the one screened exclusively with respect to the initial measurement, of A. Assertions concerning the intermediate observables $B, C, \ldots$ may be obtained readily by the standard techniques of contingent probabilities. For these frequencies we get the expression

$$P(b_k, c_m, \ldots) = D^{-1} \langle f| \ldots \rangle \ldots \langle c_m| b_k \rangle \langle b_k| a \rangle \langle a| b_k \rangle \ldots \langle \ldots |f \rangle,$$

$$D = \sum_{k'} \sum_{m'} \ldots \langle f| \ldots \rangle \ldots \langle c_{m'}| b_{k'} \rangle \langle b_{k'}| a \rangle \langle a| b_{k'} \rangle \ldots \langle \ldots |f \rangle. \tag{3}$$

This expression is not multilinear in terms of the kets and bras corresponding to the possible outcomes of the intermediate observations, because of the denominator. Nor can it be written as the product of two factors one of which depends only on the screening procedures, the other only on the observations made on the resulting ensemble. Hence there is no simple manner in which some expression may be considered to be fully descriptive of the ensemble that has been constructed.

Because of the importance of the ensemble resulting from multiple-time screening for the argument of this paper, it should be emphasized that such ensembles are not without interest in experimental work. Particularly in high-energy and elementary particle research, samples for further analysis are frequently selected, by human scanning or by fully automated procedures (e. g. master pulse techniques), with respect to characteristics of both initial and final situations.

Ensembles constructed by two-time screening are not theoretically inferior to those obtained by the more conventional procedures of pre-selection. Ordinary quantum theory furnishes probabilistic formulas that permit one to make assertions concerning the outcome of experiments performed on these ensembles. But in some important respects they have properties not shared by pre-selected ensembles. Consider, for instance, a two-time selected ensemble with two intermediate observations. If the observables corresponding to these intermediate observations do not commute with each other, it does not automatically follow that the product of their respective dispersions obeys HEISENBERG's uncertainty relation. Let, as an extreme example, the two intermediate observables, B and C, be identical with A and F, respectively, with respect to which the ensemble was screened,

$$B \equiv A, \quad C \equiv F, \quad [B, C] \neq 0. \tag{4}$$

The value of B will then be found to equal a, and that of C to be f, for all members of the ensemble. Hence the dispersion of the observations on B alone, and of those on C alone, will separately vanish, in apparent contradiction to the HEISENBERG relations. The real import of this result is, of course, not a logical contradiction but merely an injunction to be circumspect in the formulation of the uncertainty relations, which hold but for a limited class of ensembles.

I shall now return to the various roles that are played by the state vector of a quantum theoretical ensemble. First, it is part of the characterization of a type of physical system that one must construct the particular Hilbert space in which its observables are defined formally as Hermitian operators. This Hilbert space, being a linear vector space, defines a class of mathematical objects that are its constituent vectors, regardless of what additional physical meanings may be assigned to them. The present investigation changes nothing in this respect.

In the Schrödinger picture the kets are also the carriers of the dynamical law of the physical system; this role is not sacrosanct, considering the ease with which the dynamics can be fastened on the observables, in the Heisenberg picture, which not only resembles in this respect more closely the approach of classical mechanics but which, according to P. A. M. DIRAC[1], is preferable at least in some respects in quantum electrodynamics.

There remains the importance of the state of a system as a particularizing characteristic that sets it apart from other systems possessing the same structure (i. e. dynamical law) but happening to be "in a different state". This aspect of the state vector is shared by the Schrödinger and Heisenberg pictures as well as by other intermediate pictures. That the state vector has this function, the function carried in classical mechanics by the representative point in phase space (i. e. GIBBS' "phase"), is suggested by the circumstance that into the predictions concerning conventionally constructed ensembles the state vector of the systems enters as the only particularizing piece of information that is independent of the choice of the measurements to be performed. Surely, the state

---

[1] DIRAC, P. A. M.: Several public lectures, 1965 and 1966.

vector, or its generalization, the density matrix, are appropriate means for the characterization of ensembles that have been constructed by means of one-time screening prior to the onset of measurements.

For ensembles constructed by means of terminal one-time screening the state vector also is a proper means of characterization, and the only piece of information required for retrodiction. We have pointed out in Ref. 1 that in this case the ensemble is in the "state" established by subsequent screening, whereas in the conventional ensemble the state depends on the preceding screening. Two-time screening confronts us not with mere ambiguity; these ensembles simply cannot be described in terms of any one state vector or density matrix. This negative result tends in my opinion to undermine the significance of the state vector for the individual system as well.

The assignment of a state vector to a physical system operationally implies no more than the prediction of frequencies of outcomes of observations to be made subsequent to the establishment of a specified state. These predictions then refer not to individual physical systems but to ensembles composed of such systems, and moreover only to the class of ensembles constructed by means of one-time screening. Of course, ensembles are not objects occurring in nature but constructs of our intellect; what occurs in nature are individual physical systems. Systems may be imbedded in conceptual ensembles, or sets of real systems may be combined for purposes of discussion into ensembles. Experimental verification of the statistical assertions of quantum theory is based on the latter procedure. Because of the specialized character of one-time screened ensembles it appears open to question whether a means of characterizing such ensembles should be considered an intrinsic property of the component physical systems.

If the assignment of quantum states to individual physical systems is abandoned, the problematics of the change of quantum state as a result of the performance of measurements, i. e. the so-called collapse of the wave packet, evaporates. The performance of a measurement, and the use of its outcome as a means for additional screening, replaces the original ensemble by a subensemble, with its own statistical dispersion. The apparent resolution of the problems surrounding the extradynamic change in the quantum state of a system should, however, not be considered an unmitigated triumph. This advance is achieved only at the price of having to give up what hitherto has been considered the principal particularizing property of a quantum-theoretical physical system; no substitute is in sight.

Certainly, I do not consider that the discussion presented here leads to any definite conclusions. I believe that it calls attention to a possible novel point of view, whose import remains to be assessed.

Chapter 4

Probabilities in Quantum Mechanics[1]

Henry Margenau and **Leon Cohen**[2]

Department of Physics, Yale University, New Haven,
Connecticut, U.S.A.

This note seeks to survey crucial issues and to pinpoint difficulties in the current controversy over the meaning of measurement [1] and its proper mathematical description in quantum physics. It deals primarily with those philosophic and mathematical considerations which arise from the statistical nature of the results of physical measurements. The first part reviews and aims to clarify matters of more or less general interest; the second and third are devoted to some specific recent results concerning joint probabilities and phase-space distribution functions.

I. One of the important features of quantum mechanical description is the scattering, the dispersion of measured values in repeated observations of a given quantity even when the physical state (ψ-function) of the system, which is the carrier of the quantity, is as precise and determinate as human ability permits. This fact has occasioned a variety of philosophical explanations: Bohr's and Heisenberg's early belief that the statistical dispersion inherent in a quantum mechanical state is the reflection of the uncertainties introduced by the necessary interaction with the measuring apparatus which occurs at the time of measurement; De Broglie's and Bohm's suggestion that obscure factors not appearing in the analysis (hidden variables) account for these fluctuations; Heisenberg's later appeal to the Aristotelean theory of potentia [2], according to which the measurement of an observable in a state other than its eigenstate converts a potential quantity into an actually existing one; the distinction between "possessed" and "latent" observables [3], which formulates a new version of the philosophic contrast between primary and second ary qualities.

[1] Work supported by the United States Air Force Office of Scientific Research under Grant AFOSR 843/65.
[2] Now at Hunter College of the City University of New York.

The scattering of observations, and hence the use of probabilities, is of course not confined to quantum mechanics. Even in classical physics, analysis of the measurement process is meaningless without a theory of errors. The distinction between the two fields is thought to reside in the circumstance that the probabilities of classical physics are "reducible" [4], whereas those of quantum mechanics are not. That is to say, classical physics is in principle compatible with the assumption that knowledge of a system's state can be so refined that every conceivable physical quantity has an exact value, one which recurs in every measurement, so that the probability of the occurrence of any namable value is "reduced" either to zero or to one.

Max Born [5] has advanced an interesting argument which makes this statement a little artificial. Even in classical physics, he insists, the use of precise values of observables, unencumbered by probabilities, involves an idealization which defeats itself. It is surely not proper to assume exact knowledge of the instantaneous positions and velocities of all the molecules of a gas. A more reasonable premise is to assign to the values of these quantities small errors. This admission, however, opens the floodgates to ignorance, which develops in time. For the error attaching to position and velocity of a molecule increases in every collision. Roughly speaking, the relative error is doubled, and it is not difficult to see how long it will take for the error to grow to the magnitude of the quantity itself. If we start with an implausibly small error of one millionth of 1 % in the knowledge of the original velocities, our knowledge will have been converted to ignorance in much less than one microsecond for a gas under standard conditions. In general, what physicists call relaxation time is also the time required for knowledge to be converted into ignorance in problems of this sort.

Such reasoning is of great practical importance, but it does not change the fact that classical physics, with its use of precise states, remains internally consistent and allows probabilities to be in principle reduced to certainties. Even the argument just cited is predicated upon the *functioning of exact laws* controlling the evolution of exact mechanical states whose incidental description or knowledge transcends our competence. No inherent principle of nature prevents us from ascertaining as exact a value as we please of the position and velocity of a single molecule at a given time, and there are ways in which we can study its behavior during an interval much smaller than the relaxation time, so that a dynamic account of its motion is possible. This leaves the distinction between quantum mechanics and classical physics as profound as ever; nor was it the intention of Born, whose genius recognized it before all others, to deny it. His example serves to show forth the universal importance of probability.

Measurement is always a statistical affair. No single observation can be trusted to yield what goes as the "true" value of a physical quantity. Perhaps it is not amiss to sketch briefly how that "true" value is established.

Suppose we are making N measurements of a given quantity with the same apparatus. If N is sufficiently large, these results enable us to construct a curve, usually a Gauss error curve, which has a certain maximum and a certain half width. The half width is related to the probable error of the set of measurements, and the maximum of the curve defines the true value of the quantity. If the curve is indeed an error curve, then it may be shown that the arithmetical average of all the measured values are equal to the maximum of the curve. If N increases, the curve retains its width but it is possible to place the points more accurately upon it and thereby to define the maximum, the true value, more precisely. Hence, the probable error remains no matter how large the value of N.

To reduce the width of the curve different measuring apparatus must be employed and the more precise the measuring device the narrower the curve which results from the measurements. Thus arises the question: Is it possible within reasonable limits to choose different measuring instruments of increasing precision in such a way that the width of the curve becomes smaller and smaller but falls each time within the range of the preceeding grosser curve? For instance if we are to measure the length of a given object, the first device might be a carpenter's rule, the second a carefully calibrated yardstick, the third a Vernier caliper, the fourth a traveling microscope, the fifth an interferometer device, and so forth. In classical physics it had always been assumed that this kind of convergence, the nestling of successive error curves of smaller and smaller widths within the larger, was a fact of nature. Quantum mechanics revealed that there exists no set of measuring apparatus of progressive refinement which will make this kind of convergence true. In the fine-grain domain of atomic physics, errors cannot in general be confined to preassigned limits, and this is because of HEISENBERG's indeterminacy principle, which we shall now briefly consider.

In the form $\Delta p \cdot \Delta q \geq \hbar/2$ the uncertainty principle is said to relate the "error" in the measured momentum p to the "error" in the position q of a particle. Physicists understand the application of this inequality very well and employ it without difficulty, but the explanation they give of it, especially in textbooks, is often highly questionable. Sometimes, major atrocities are committed in its name.

The most prevalent simple view concerning the origin of uncertainty in quantum mechanics holds the disturbance during the measurement process responsible for it. The γ-ray microscope carries much persuasive force in this line of reasoning: To measure q accurately one needs a short wave length γ-ray, and this produces, by way of the particle's recoil, an uncontrollable and unpredictable effect upon its momentum; hence a q-measurement entails an unavoidable error in p, and vice versa. Now all this, and the neat algebra that usually goes with the discussion, are certainly correct, but what they have to do with HEISENBERG'S principle is far from clear. *Four very elementary difficulties emerge at once.*

The inequality asserts that in any state of the particle that is at any instant of time, and whether a measurement is being performed or not, the uncertainties or errors, cannot be smaller than it specifies. According to the disturbance theory, however, where Δp is due to the measurement interaction, it can only refer to the state of the particle *after* the measurement, while Δq relates to the position *before*. To contradict this inference would be to hold that measurements do not reveal what is, but what will be, or that they can play either role.

The second difficulty lies in the use of words like "unpredictable" and "uncontrollable". The disturbance theory has a decidedly classical cast; it pursues the interaction of the particle with the photon in complete classical detail, and if it were consistent it should admit both the controllability and the predicability of the results, for they are provided by electrodynamics. But here, as a concession which completely surrenders the point at issue, it interposes a hiatus and brands the interaction mysterious. The argument therefore begs the question.

Thirdly, the disturbance theory — unless it is taken as an inductive exemplification or merely as an illustration of the uncertainty principle — violates logic. Surely, quantum mechanics is a wider discipline than classical physics, based upon axioms additional to those which underlie the latter science. In other words, classical physics is a limiting case of quantum mechanics. Hence it can not possibly contain it logically since its range is smaller, and one cannot use classical reasoning to derive the uncertainty principle.

Finally, there is trouble with the definition of the Δ symbol. Errors, as we have seen, have no meaning with respect to a single measurement; being deviations from a mean they necessarily presuppose an ensemble of values. Thus, unless the γ ray microscope story is extended to embrace a great number of interactions between a particle and a photon and Δ is understood to refer to the statistical deviation of the results, the disturbance hypothesis is without relevance. If it is so extended, then it becomes unclear why only in the majority of instances, and not in all, an energetic photon causes a large disturbance, for there are results for which the deviation from the mean is very small. To us there appears no way in which all four difficulties can be removed.

Besides the disturbance theory, there is another popular simple conjecture which is invoked to explain uncertainty, again within the frame of the conventional pictorial concepts of mechanistic physics. Here the approach is operational; it is said, and made plausible by numerous examples, that it is instrumentally impossible to measure p and q simultaneously; the hardware clashes. This claim is expanded to cover all non-commuting observables: failure of operators to commute is supposed to symbolize experimental incompatibility of measuring procedures. When coupled with the doctrine of disturbance, the claim renders a rather pleasing account, for if p must invariably be measured before or after q, the effect of the interference can not be avoided.

The outlawing of simultaneous measurements, however, entrains its own peculiar infelicities. In the first place, there is no consideration whatever which prevents an experimenter from letting a particle reflect, at the same time or as nearly the same time as he pleases, a soft X-ray and a hard γ-ray. He can thus obtain numbers to be assigned to p and q, and this belies the thesis barring simultaneous measurements. To be sure, the simultaneous values for p and q thus obtained — whose errors escape our knowledge because there is no ensemble — can not be used as premises for causal prediction, they are without interest by themselves, and they do not define at quantum mechanical state. But they are available, and this makes the claim under examination palpably false.

But it exhibits a further, even more serious philosophic flaw. Science is the rational accomodation of contingent facts or observations. A theory which prescribes what can in the future be achieved or, worse, proscribes specific empirical acts, like simultaneous observations of p and q, transcends its own competence by restricting the development of experimental techniques. It becomes legislative when it should be explanatory. For these reasons it is necessary to look deeper into the bases of indeterminacy, and this forces us to renounce the encrusted habits of visual pictorization of elementary events, to relinquish the attempt to explain quantum uncertainty in terms of the familiar notions of conventional particle trajectories or wave propagation. Although it is heresy to say so, we believe that there is no dualism, no complementarity in quantum physics. The electron is neither a particle nor a wave, even if our accustomed language induces us to use these words. They are, strictly speaking, metaphorical and allude to something whose description exceeds the bounds of visual perception and is therefore obliged to discard such concepts as particles and waves, except as imperfect models of reality. Classical physics developed in the aegis of Cartesian clarity, its concepts answered the question: What can the mind's eye imagine? Quantum mechanics was born out of concerns characterized by the question: What can be measured? It has now left this area and has begum to ask: What can be conceived abstractly, but with mathematical and logical consistency?

After this digression we return to a correct, if minimal, interpretation of the uncertainty principle, which, though short of serving as an explanation, goes far in showing the inadequacy of the foregoing views. If one follows through the mathematical derivation of the formula, $\Delta p \cdot \Delta q \geqq \hbar/2$, the meaning of Δp and Δq becomes very clear; they represent the standard derivations, or the square root of the variances, of sets of repeated observations of p and q when the system (loosely speaking, the particle) is in an arbitrary but definite quantum state ψ. For example, we prepare the state of the system at time t, by passing it through a filter, or through a Stern-Gerlach field, or merely by waiting until an atoms has settled to its lowest energy state. Then, at time $t_1 + \tau$, we perform a measurement of p, obtaining a value p_1. What happens to the system after $t_1 + \tau$ depends on the manner in which the measurement was made, and knowledge of ψ does not in general permit its prediction. Different interactions may change ψ in many different ways; what is important from the point of view of the measurement is the number p_1.

To procedure an ensemble, the system must be reprepared at a later time t_2. A second measurement of p is then performed at $t_2 + \tau$, and the result is p_2. In this way, by m repreparations and measurements, we obtain the set of numbers $p_1, p_2 \ldots p_m$. By a corresponding procedure we can generate a set $q_1 \ldots q_n$. The principle asserts that the standard deviations of these p's and these q's satisfy the inequality, for any ψ.

From this discussion it becomes apparent that the "error" in q, or Δq, can not be caused by the disturbance wrought in the measurement of any p, for the system was in each instance reprepared. Nor does it matter in what order the values appear, or in what manner the q-measurements are interspersed among those of p.

A succession of repeated preparations creates an ensemble, the instances of which are formed by one (perhaps the same, although this can rarely be guaranteed) system in the same state at different times. One might call this a time ensemble. Quantum mechanics is known to apply to space ensembles too. These are present when many systems, each in the same state, and isolated from each other, are present at the same time and the measuring device interacts with them all. A single observation will then yield the set $p_1 \ldots p_m$, where m is now the number of systems present. Another single observation gives $q_1 \ldots q_m$. Indeed these observations can also be performed at the same time. In practice, the space ensemble is of more common occurrence, but the uncertainty principle holds for both.

In the sequel we shall designate this just preceding account of the principle as its *standard version*. Whether it admits other interpretations is at the moment not clear.

We now examine briefly the logic of the probability concept as it is employed in quantum mechanics. As is well known, there are two extreme, basic approaches to the meaning of probability [5]. One, sometimes called a priori or subjective, establishes probability as a measure of the degree of confidence or expectation which a person may have in the outcome of an event. Its first mathematical formulation was given by LAPLACE, and one of its contemporary advocates is H. JEFFREY [6]. When viewed in this way, probability becomes in principle untestable and, at least in its most elementary sense, it changes with incidental evidence concerning the event. For instance the appearance of a number from 1 to 6 when a die is thrown is known to be $\frac{1}{6}$ before this event. If probability is lodged in one's knowledge with respect to the outcome of the next throw, its value changes from $\frac{1}{6}$ either to 1 or 0; the event "reduces" it to certainty. There are ways of escaping this conclusion even while maintaining the subjective interpretation, but they are not interesting here because it is just this conclusion, which springs from the most radical subjectivism in the logic of probability, that has left an imprint on the philosophy of quantum mechanics.

The principal representative of various objective views is the frequency theory, which holds probability to be the relative frequency of the occurrence of a specified event, or of a set of events, in a large aggregate of occurrences. The probability of throwing a 5 with a die is defined as the ratio of the number of times a 5 appears in a series of throws to the total number of throws. Although this ratio fluctuates as the number of throws increases, and indeed never attains a limit in the ordinary mathematical sense, it has become possible to define it with a precision sufficient for successful use.

It may not be out of place to emphasize here that probability in the latter sense, is a measurable quantity, entirely on a par with every other physical quantity. This needs to be recalled in order to counteract the superstition that probability stalks like a ghost through science, that its invocation is always a concession of ignorance. As a matter of fact, the assignment of probabilities to the numbers on the faces of a die, if they are understood in the objective manner, are attributes of the die which are as real as its color or its size or its weight. The reason people regard them otherwise is in the circumstance that probabilities can not be observed in one fell swoop, in a single measurement. However, as our earlier discussion shows, the other physical properties also require in essence a large number of observations in order that their "true" values be found. The psychological bias causing many to look upon probabilities as scientific citizens of second rank seems to stem from its lowly birth in minds preoccupied with games of chance.

One last point on the general logic of probability. It is wrong to view the subjective thesis as *opposed* to the objective one. Every physical theory has necessarily a purely formal aspect and an operational one. The former allows it to predict, the latter to confirm or refute its predictions [5]. The two logically disparate versions of probability here reviewed are therefore complementary to each other, and the very existence of these two required components is added evidence for the normal and respectable status of probability as a physical quantity.

Our discussion of the uncertainty principle, in its standard version, makes clear at once that the objective frequency interpretation is relevant for quantum mechanics. If the state function ψ has its postulated meaning, it too must be given a frequency sense. This has several consequences. First, since a relative frequency can never be determined by means of a single observation, since one throw of a die does not determine the next, there can be no presumption that a single observation of ψ, i. e. a measurement, *must establish its form,* or *modify it in a way fully known,* nor that is must determine the outcome of the next measurement. The other consequence is that a single observation does not change the probability in question at all. Objectively, whether or not a 2 appears in the next throw of a die, the probability of its appearance goes right on being $\frac{1}{6}$ afterwards.

Yet both of these conclusions are often violated when the meaning of ψ is discussed by physicists. von Neumann introduced the so-called projection postulate; textbooks speak of the reduction of a wave packet upon measurement; these amount to the posit that a single measurement has the effect of producing an eigenstate, namely one corresponding to the measured value. This hypothesis has been shown to be untenable elsewhere (references 4, 5 and literature cited there) and its short comings need not be exhibited again. It might be of some interest, however, to speculate upon the reasons why this strange hypothesis has had such persuasive force.

One of them is surely psychological; the projection idea is a hangover from classical physics, where every measurement (in a very naive sense which is actually contradictory to the theory of errors) insures the same outcome upon repetition. There, every measurement is also the preparation of a state. In quantum mechanics this need not be the case.

The other reason is mathematical. Quantum mechanics associates operators with physical acts yielding numbers. Measurement itself represents the most universal act of this type, and it is tempting to seek an operator, M, which symbolizes it in universal fashion. Specific operators for specific measurements are already known. Now, what are the properties of measurement in general? To say that it always generates a real number is not helpful in this search, for it yields an indiscriminate, large class of operators. One may look upon it in another way, however. A measurement puts a question to nature: does this variable have a value within this given range? The answer is either yes or no, it can be symbolized as 1 or 0. Hence the operator, M, must have two eigenvalues, M', namely 1 and 0, and no more. These satisfy the equation $M'^2 = M'$; hence M must satisfy it also. But the operator, defined by $M^2 = M$ is known to be the projection operator, constructible for any state ψ in well known ways.

This somewhat formal success of the search for M entails a mathematical conclusion. When M is applied to a vector in Hilbert space, say x, it changes the direction of x and reduces its magnitude so that it becomes the component of x in a new direction. In short, it projects x upon some ray in Hilbert space. Physically, this means it creates a new state. Hence the projection hypothesis.

The cogency of such reasoning is of course spurious. Even if M is accepted as a valid symbol for a universal measurement, the application of M to an *arbitrary* x need not be interpreted as the outcome of measurement upon a system in state x at all. There is no warrant for this hypothesis in the axioms of quantum mechanics. More serious, however, is the identification of M with measurement in general. Why should there be a valid symbol of a specific sort for every possible measurement process? In classical physics we have formulas for momentum, kinetic and potential energy and all other important observables. But do we have, or *need*, a formula for an unspecified observable? Our suspicion is that it would be trivial if written down.

We therefore end this section with the assertion that quantum mechanics deals with measurable probabilities which take the form of relative frequencies; that the interaction of a physical system with a measuring device does not necessarily project its state onto an eigenvector in Hilbert space-but rather projects thath state into irrelevance. As a rule, quantum mechanics ceases to have interest in the state of a system after an observable has been measured upon it. There is no single axiom which defines the post-measurement condition. Nevertheless, there are operations (e. g. correlation experiments) which insure that a given observed value will always be associated, simultaneously or in sequel, with a specifiable other value, observed or not. These, however, are not governed by one basic axiom and require special treatment. They are best viewed as single but compound measurements [7].

II. In section I attention has been confined to the probability that, the state of a system being ψ, a measurement of an observable r shall give the eigenvalue r_i. It is specified by the postulates of quantum mechanics as

$$P(r'; \psi) = |\langle \psi | r_i \rangle|^2 \tag{1}$$

provided $\langle \psi | r_i \rangle$ is the scalar product of the bra vector $\langle \psi |$ and the ket vector $| r_i \rangle$, which is an eigenstate of r corresponding to r_i. In the preceding section we reviewed and criticised various hypotheses relating to the measurement of two different observables, say x and y (e. g. position and momentum), simultaneously or in sequence, and therefore the question arises as to the probability that, when a quantum system is in state ψ, a measurement of x and one of y shall yield the eigenvalues x_i and y_j. The cases in which $y \equiv x$, or $y \equiv (x$ at a later time) are included here. This introduces the concept of a joint probability, to be designated by $P(x_i, y_j; \psi)$. The ordinary postulates are non-committal with respect to it construction, and here lies the mathematical root for the disparity of views regarding what happens when p and q, for instance, are being measured. Some philosophers of physics argue in fact that a construct like $P(x_i, y_j; \psi)$ is meaningless except when the operators for x and y commute, usually by an appeal to the belief that joint measurements for them are impossible. They are then inclined to devise non-Aristotelean quantum logics which open unlimited vistas for speculation. We have already given reasons to doubt this conjecture. If it were true in the literal sense of every proper theory of stochastic variables, then it *ought* to be possible to construct a joint probability, and furthermore $P(x_i, y_j; \psi)$ should be zero for every i and j when x and y do not commute. At any rate, the search for possible formulations of a joint probability function must not be foregone. If none exists, that is at least worth knowing.

The search begins with a statement of the mathematical conditions which a joint probability must satisfy. These are, positiveness of P, (2) and (3, 4) the necessary relation to the marginal probabilities. They have the form

$$P(x_i, y_j; \psi) \geq 0, \tag{2}$$

$$\sum_i P(x_i, y_j; \psi) = P(y_j, \psi) = |\langle \psi | y_j \rangle|^2, \tag{3}$$

$$\sum_j P(x_i, y_j; \psi) = P(x_i, \psi) = |\langle \psi | x_i \rangle|^2, \tag{4}$$

One may wish to impose further conditions, demanding that, when used in the calculation of expectation values in the customary way, they shall lead to the correct quantum mean; e. g. that

$$\text{Exp}(x) = \sum_{ij} P(x_i, y_j; \psi) \, x_i = \langle \psi | x | \psi \rangle,$$

where x_{op} is the operator corresponding to the observable x. This greatly encumbers the search. We return to this problem in the following section and restrict ourselves here to conditions 2, 3 and 4 which seem to suffice as a basis for a minimal theory of measurement.

A theorem of statistics relates the covariance of two random variables x and y to their joint probabilities in the following way.

$$\text{Cov}(x\,y) = \sum_{ij} P(x_i, y_j)\, x_i\, y_j - \langle x \rangle \langle y \rangle. \tag{5}$$

As to notation, the covariance

$$\text{Cov}(x\,y) \equiv \langle x\,y \rangle - \langle x \rangle \langle y \rangle$$

provided we use brackets to denote expectation values. Now quantum mechanics immediately suggests that we write for $\langle x\,y \rangle$ something like $\langle \psi |\, x\,y\, | \psi \rangle$, x and y being the operators corresponding to x and y, and if this is done one can extract the form of $P(x_i, y_j, \psi)$ from (5). Certainly $\langle x \rangle$ and $\langle y \rangle$ are known to be $\langle \psi |\, x\, | \psi \rangle$ and $\langle \psi |\, y\, | \psi \rangle$.

But the difficulty encountered here arises from the fact that the product $x\,y$ can be translated into quantum mechanics in many different ways. What is needed is a *rule of correspondence* between a product like $x\,y$, or in general $x^n\,p^m$, in which the factors commute and which is directly observable, and its quantum equivalent. Here appears in very specific form the fundamental epistemological problem of the relation between direct experiences (*P*-plane of reference 5) and the constructs that symbolize them in our reasoning about the world (*C*-field). In classical physics that relation was held to be a trivial isomorphism — the construct temperature *was* the class of numbers provided by a thermo-meter; however here the rules of correspondence become a matter for deliberate choice and discrimination. More will be said about them in part III.

Suffice it here to record that a fairly obvious choice for $x\,y$ is obtained by symmetrization [8]:

$$x\,y \to \frac{x\,y + y\,x}{2}.$$

This leads, via Eq. (5) to the result for the joint probability

$$P(x_i, y_j; \psi) = R\,[\langle \psi | x_i \rangle \langle x_i | y_j \rangle \langle y_j | \psi \rangle] \tag{6}$$

where R stands for "real part". Formula (6) has certain attractive features. It provides a correlation coefficient $\sigma(x, y)$ defined by

$$\sigma(x, y) = \text{Cov}(x\,y)\,[\text{Var}(x)\,\text{Var}(y)]^{-\frac{1}{2}} \tag{7}$$

which satisfies the desired inequality

$$-1 \leq \sigma(x\,y) \leq +1$$

for every x and y. It also conforms to Eqs. (3) and (4), but unfortunately it does not satisfy (2). P constructed in accordance with (6) is not in general positive definite, as closer inspection (and various examples worked out in reference 8) will show.

This one example, to which others will be added in the final section of this paper, places in view the major difficulty which afflicts nearly all formulations of $P(x_i, y_j; \psi)$. There may be a deeper reason for that, possibly hidden in strange and obscure features of the measurement process which make it similar to creation and annihilation in field theories[1]. At the moment, however, we shall take the stand that joint probabilities which violate (2) are to be rejected.

One then observes that there is *one* $P(x_i, y_j; \psi)$, a trivial one, which does obey all sum rules, (2—4), and which can be written down even without the detour over covariances. It is

$$P(x_i, y_j; \psi) = \left|\langle \psi | x_i \rangle\right|^2 \left|\langle \psi | y_j \rangle\right|^2. \tag{8}$$

While mathematically adequate from the present point of view, it has the fault of yielding no correlations between the random variables x and y: $\mathrm{Cov}(xy)$, constructed in conformity with (5) is zero, and so is $\sigma(x, y)$. Must one therefore reject it?

This may well be the case, but for reasons that are not yet wholly clear. It is noteworthy, however, that formula (5) contains exactly the proposition which was affirmed in our discussion of the standard version of uncertainty. There we saw that the probability of measuring p, *after repreparation of the state* on the time ensemble *or in simultaneous observations* on *a space ensemble*, was indeed independent of the measurement of q. Formula (8) describes that situation. It seems, therefore, that we are in possession of a reasonable theory of measurement, which endows joint probabilities with meaning, provided we accept the conditions noted in part I; they are: an interpretation of uncertainty as standard deviation of a set of measured values after repeated preparation of the state ψ, and the disavowal of a priori knowledge about the fate of the quantum system after measurement.

We now turn to the consideration of a special class of joint distribution functions, namely $P(q, p; \psi)$, where q and p continue to signify position and momentum. These are phase space distribution functions. Through ψ, the functional P depends also on the time t. In the next section we employ a notation which lends itself a little more naturally to the technical matters under discussion; we write

$$P\big(q, p; \psi(t)\big) = F(q, p, t)$$

---

[1] This point was emphasized to us in conversation by Professor Vigier.

omitting the argument t when the time dependence is otherwise clear. Only one-dimensional systems will be treated; generalization to several dimensions is easy but entails more cumbersome equations. Finally, use of the ordinary notation, $\psi(q, t)$ in place of $\langle\psi| q, t\rangle$, is simpler and will therefore be adopted.

III. The possibility of formulating quantum mechanics in terms of ensembles in the phase space of position and momentum originated with WIGNER [9]. He found a function of position and momentum which satisfies Eqs. (3) and (4) for the case of q and p. The Wigner distribution is

$$F_w(q, p) = \frac{1}{2\pi} \int \psi^*\left(q - \frac{1}{2}\tau\hbar\right) e^{-i\tau p}\, \psi\left(q + \frac{1}{2}\tau\hbar\right) d\tau. \qquad (9)$$

It can readily be verified that

$$\int F(q, p)\, dp = |\psi(q)|^2, \qquad (10)$$

$$\int F(q, p)\, dq = |\varphi(p)|^2 \qquad (11)$$

where $\varphi(p)$ is the momentum state function given by

$$\varphi(p) = \frac{1}{\sqrt{2\pi\hbar}} \int \psi(q)\, e^{-i/\hbar\, qp}\, dq \qquad (12)$$

and $|\varphi(p)|^2$ the quantum mechanical probability distribution for momentum. Although F_w does yield the correct marginal distributions, one cannot say that F_w is a true joint distribution of q and p because it does not in general satisfy (2). None the less, F_w can be formally used, to some extent, to calculate expectation values of quantum observables using phase space integration. A different distribution function (studied by MARGENAU and HILL [8]) is

$$F_1(q, p) = \frac{1}{4\pi} R\left\{\psi(q) \int e^{-i\tau q}\, \psi^*(q - \tau\hbar)\, d\tau\right\} \qquad (13)$$

where R signifies the real part of the quantity in brackets. This represents another way of writing Eq. (6). If one assumes that position and momentum are not correlated, then we are lead to take distribution (8) which now reads

$$F_0(q, p) = |\psi(q)|^2\, |\varphi(p)|^2. \qquad (14)$$

Both (13) and (14) satisfy (10) and (11).

An explicit expression for the totality of functions of q and p which satisfy (10) and (11) can be given [10, 11]. It is

$$F(q, p, t; f) = \frac{1}{4\pi^2} \iiint e^{-i\vartheta q - i\tau p + i\vartheta u} \times$$
$$f(\vartheta, \tau, t)\, \psi^*\left(u - \frac{1}{2}\tau\hbar\right) \psi\left(u + \frac{1}{2}\tau\hbar\right) d\vartheta\, d\tau\, du \qquad (15)$$

6*

where $f(\vartheta, \tau, t)$ is any function which satisfies

$$f(0, \tau, t) = f(\vartheta, 0, t) = 1. \tag{16}$$

Thus f is the key to an indefinite number of phase space distribution functions which can be generated via Eq. (15). That (15) is the set of all functions constrained by conditions (10) and (11) can be proved by first showing that a judicious choice of f will allow any function of q and p to be written in form (15)[1]. Then the integration of F with respect to p and q forces the imposition of (16) on f if the result of the integration is to yield (10) and (11). F_w, F_1 and F_0 are obtained from (15) by particularizing f to the value

$$f_w = 1 \tag{17}$$

$$f_1 = \cos \tfrac{1}{2} \vartheta \, \tau \, \hbar \tag{18}$$

$$f_0 = \frac{\iint |\psi(q)|^2 \, |\varphi(p)|^2 \, e^{i\vartheta q + i\tau p} \, dq \, dp}{\int \psi^*(u - \tfrac{1}{2}\tau\hbar) \, e^{i\vartheta u} \psi(u + \tfrac{1}{2}\tau\hbar) \, du} \tag{19}$$

respectively.

The transition from quantum to classical mechanics can be made very clear through the investigation of the time derivative of F, i. e. the "equation of motion" which F satisfies. If (15) is differentiated with respect to time and Schrödinger's equation is applied to ψ, one obtains, after a somewhat lengthy but direct calculation, the equation of motion for F.

$$
\begin{aligned}
\frac{\partial F(q, p, t; f)}{\partial t} = {} & \frac{\dot{f}\left(-i\dfrac{\partial}{\partial q_F}, -i\dfrac{\partial}{\partial p_F}\right)}{f\left(-i\dfrac{\partial}{\partial q_F}, -i\dfrac{\partial}{\partial p_F}\right)} F(q, p, t; f) \\
& + \frac{2}{\hbar} f^{-1}\left(i\frac{\partial}{\partial q_F}, i\frac{\partial}{\partial p_F}\right) f\left(-i\frac{\partial}{\partial q_H}, -i\frac{\partial}{\partial p_H}\right) \times \\
& \times \sin \frac{\hbar}{2}\left[\frac{\partial}{\partial p_F}\frac{\partial}{\partial q_H} - \frac{\partial}{\partial p_H}\frac{\partial}{\partial q_F}\right] \times \\
& \times f\left(i\frac{\partial}{\partial q_F} + i\frac{\partial}{\partial q_H}, i\frac{\partial}{\partial p_F} + i\frac{\partial}{\partial p_H}\right) H(q, p)\, F(q, p, t; f)
\end{aligned}
\tag{20}
$$

where $\dfrac{\partial}{\partial q_H}, \dfrac{\partial}{\partial p_H}$ operate on H only and $\dfrac{\partial}{\partial q_F}, \dfrac{\partial}{\partial p_F}$ on F. $H(q, p)$ is the *classical* Hamiltonian. Now, suppose we choose f as a function of $\hbar$

---

[1] By taking the Fourier transform of (15) we have

$$f(\vartheta, \tau, t) = \frac{\iint F(q, p, t)\, e^{i\vartheta q + i\tau p} \, dq \, dp}{\int \psi^*(u - \tfrac{1}{2}\tau\hbar, t)\, e^{i\vartheta u} \psi(u + \tfrac{1}{2}\tau\hbar, t)\, du}.$$

Thus for any well-behaved F, we can find an f so that (15) is satisfied. We must, of course, also know φ.

such that

$$\lim_{\hbar \to 0} f \to 1, \tag{21}$$

$$\lim_{\hbar \to 0} \dot{f} \to 0. \tag{22}$$

Since

$$\sin \frac{\hbar}{2} (\ ) \to \frac{\hbar}{2} (\ ),$$

Eq. (20) becomes, in the limit as $\hbar$ approaches zero,

$$\frac{\partial F}{\partial t} = \frac{\partial H}{\partial q} \frac{\partial F}{\partial p} - \frac{\partial H}{\partial p} \frac{\partial F}{\partial q} \tag{23}$$

which is LIOUVILLE's equation. The Wigner distribution and the distribution of MARGENAU and HILL satisfy (21) and (22). But the distribution which shows no correlations, that is defined in Eq. (14) which resulst from

$$f(\vartheta, \tau) = \frac{\iint |\psi(q)|^2 \, |\varphi(p)|^2 \, e^{i\vartheta q + i\tau p} \, dq \, dp}{\int \psi^*(u - \frac{1}{2}\tau\hbar) \, e^{i\vartheta u} \psi(u + \frac{1}{2}\tau\hbar) \, du} \tag{24}$$

does not in general satisfy (21) and (22). Hence F_0 does not obey the Liouville equation in the limit of $\hbar \to 0$. The reason for this failure is the fact that the Liouville equation reflects the presence of correlations while (14) admits no correlations in any limit.

As we have seen, the formation of the quantum mechanical operator from its classical counterpart is straightforward as long as the classical quantity is either a function of q or of p only, or if it is the sum of such functions. One merely replaces p by the usual operator p and q by q. But if the classical observable contains product terms of q and p, then difficulties arise, for there is no unique way of forming the quantum mechanical operator. Several methods have been proposed for dealing with such cases. They are called rules of correspondence, and they were encountered and commented on in section II. The following rules are known and have been used.

a) DIRAC's rule:

$$\{\mathscr{A}, \mathscr{B}\} \to -\frac{i}{\hbar} [\boldsymbol{A}, \boldsymbol{B}] \tag{25}$$

where $\{\ ,\ \}$ is the classical Poisson bracket of $\mathscr{A}$ and $\mathscr{B}$ and $[\ ,\ ]$ is the commutator of the operators $\boldsymbol{A}$ and $\boldsymbol{B}$

b) VON NEUMANN's rule:

If

$$\mathscr{A} \to \boldsymbol{A}$$

than for any function g

$$g(\mathscr{A}) \to g(\boldsymbol{A})$$

and if

$$\mathscr{A} \to A$$
$$\mathscr{B} \to B$$

then

$$\mathscr{A} + \mathscr{B} \to A + B$$

c) Weyl's rule:

$$q^n\, p^m \to \frac{1}{2^n} \sum_{l=0}^{n} \binom{n}{l} q^{n-l}\, p^m\, q^l.$$

d) Rule of symmetrization:

$$q^n p^m \to \frac{1}{2} \left(q^n\, p^m + p^m\, q^n \right)$$

e) Rule of Born and Jordan:

$$q^n\, p^m \to \frac{1}{m+1} \sum_{l=0}^{m} p^{m-l}\, q^n\, p^l.$$

The first two rules have been shown to be inconsistent [12].

There is an intimate connection between correspondence rules and the distribution functions F. As noted, this is due to the fact that a correspondence rule enables the calculation of the moments $\langle q^n\, p^m \rangle$, which in turn are generally sufficient to calculate the distribution function[1]. Hence we expect that for each distribution function given by (15), we can obtain a rule of correspondence. This is indeed the case. If $g(q, p)$ is the classical function, then it can be shown that the quantum mechanical operator, $G(q, p)$ is given by

$$G(q, p) = \iint \gamma(\vartheta, \tau)\, f(\vartheta, \tau)\, e^{i\vartheta q + i\tau p}\, d\vartheta\, d\tau \tag{26}$$

where $\gamma(\vartheta, \tau)$ is the Fourier transform of $g(q, p)$ and $f(\vartheta, \tau)$ satisfies (16). An additional condition which must be imposed on f to assure that G is hermitian is

$$f(\vartheta, \tau) = f^*(-\vartheta, -\tau). \tag{27}$$

Proof of these statements can be found in references 10 and 11.

---

[1] This is done by forming the so-called characteristic function $M(\vartheta, \tau)$ from the moments,

$$M(\vartheta, \tau) = \sum_{n, m=0}^{\infty} \frac{(i\vartheta)^n (i\tau)^m}{n!\, m!} \langle q^n\, p^m \rangle.$$

The distribution function, F, is then

$$\frac{1}{4\pi^2} \iint M(\vartheta, \tau)\, e^{-i\vartheta q - i\tau p}\, d\vartheta\, d\tau.$$

See any standard text in probability theory.

Different choices of f yield different correspondence rules. Actually (26) generates all possible correspondence rules in the sense that once we have decided that

$$q^n \rightarrow q^n \tag{28}$$

and

$$p^n \rightarrow p^n \tag{29}$$

for all n, Eq. (26) gives the only possible choices of G which reduce to (28) or (29) when $g(q, p) = p^n$ or q^n. The Weyl rule, or the symmetrization rule, and the rule of BORN and JORDAN can be obtained from (26) by taking

$$f = 1, \quad \cos \frac{1}{2} \vartheta \tau \hbar, \quad \frac{\sin \frac{1}{2} \vartheta \tau \hbar}{\frac{1}{2} \vartheta \tau \hbar}$$

respectively. The choice of

$$f = \frac{\sin \frac{1}{2} \vartheta \tau \hbar}{\frac{1}{2} \vartheta \tau \hbar}$$

when substituted in (15) yields the distribution function

$$F(q, p) = \frac{2}{\hbar} \frac{1}{4 \pi^2} \iiint \frac{e^{-i\tau p - i\vartheta q + i\vartheta u}}{\vartheta \tau}$$
$$\sin \tfrac{1}{2} \vartheta \tau \hbar \cdot \psi^*(u - \tfrac{1}{2} \tau \hbar) \psi(u + \tfrac{1}{2} \tau \hbar) \, d\vartheta \, d\tau \, du. \tag{30}$$

Having at our disposal the set of all possible correspondence rules, it is natural to ask whether any one of them can be applied in a consistent manner. By this we mean the following: In quantum mechanics the operator which represents a function, say K, of the operator $G(q, p)$ is $K\left(G(q, p)\right)$. Classically the observable which represents a function of $g(q, p)$ is also $K\left(g(q, p)\right)$. Now, is it possible to find a rule of correspondence such that the same rule can be used not only to obtain $G(q, p)$ from $g(q, p)$ but also $K\left(G(q, p)\right)$ from $K\left(g(q, p)\right)$? The following discussion will show that the answer to this question is no.

The question is tantamount to asking: can quantum mechanics be formulated as a normal stochastic theory? The fact that Eq. (15) generates all distribution functions which satisfy (10) and (11) leads us to consider the possibility of formulating quantum mechanics as a classical stochastic theory. If this could be accomplished then some of the "paradoxical", or better, novel statements of quantum mechanics could be cast in more conventional language with the possibility of their resolution in terms of classical concepts. Let us list the requirements which the distribution function, $F(q, p, t)$, must satisfy if quantum mechanics is to be cast as a classical probability theory.

They must include the conditions 2, 3 and 4 which served as the basis for our discussion of measurement theory, but they also require that the formation shall yield the correct quantum mechanical expectation values for all observables. We therefore make the following more stringent demands.

a) Since the distribution F should be a probability function, it must be non-negative definite for all values of q and p.

b) The distribution function must yield the correct quantum mechanical marginal distribution functions, as before.

$$\int F(q, p)\, dp = |\psi(q)|^2,$$
$$\int F(q, p)\, dq = |\varphi(p)|^2.$$

As these were the conditions used to built up the set of functions given by Eq. (15), we automatically know that any F does indeed yield the marginal distributions, provided condition (16) is satisfied by $f(\vartheta, \tau)$.

c) The expectation values of observables obtained through phase space integration, using F, must give the same results as would be obtained by use of the quantum mechanical methods. That is, if $g(q, p)$ is the classical function to which corresponds the quantum mechanical operator $G(\boldsymbol{q}, \boldsymbol{p})$, then we require that

$$c_1) \qquad \langle \psi | \, G(\boldsymbol{q}, \boldsymbol{p}) \, | \psi \rangle = \iint g(q, p)\, F(q, p)\, dq\, dp$$

and also that, for any function K,

$$c_2) \qquad \langle \psi | \, K\left(G(\boldsymbol{q}, \boldsymbol{p})\right) | \psi \rangle = \iint K\left(g(q, p)\right) F(q, p)\, dq\, dp.$$

Condition c_1) can always be satisfied. But once an F is chosen to satisfy c_1), this same F cannot be used to calculate the expectation value of $K\left(g(q, p)\right)$. Hence c_2) cannot in general be satisfied of c_1) is true. The proof can be found in (11).

These results also answer the question we raised. For it is clear that if it were possible to find an f such that

$$g(q, p) \rightarrow G(\boldsymbol{q}, \boldsymbol{p})$$
and
$$K\left(g(q, p)\right) \rightarrow K\left(G(\boldsymbol{q}, \boldsymbol{p})\right)$$

then c_1) and c_2) would be compatible for the same f.

We believe that these considerations, and the proof to which we have referred, definitively settle a question which has been asked and answered inconclusively in many different ways. Our answer does not, however, imply that no joint probability distribution exists which is compatible with the uncertainty principle for position and momentum. For example, F_0 of Eq. (14), meets all the requirements of a probability distribution, and yet the uncertainty principle does follow from it [8].

References

[1] ALBERTSEN, J.: Phys. Rev. **129**, 940 (1963). — ARAKI, H., and M. YANASEE: Phys. Rev. **120**, 622 (1961). — DURAND, III. L.: Phil. Sci. **27**, (2) (1960). — HEISENBERG, W.: La nature dans la physique contemporaine. Paris: Gallimard 1962; — Trad. de „Das Naturbild der heutigen Physik". Hamburg: Rowolt 1955. — JAUCH, J. M.: Helv. Phys. Acta **33**, 711 (1960). — JAUCH, J. M., and G. PIRON: Helv. Phys. Acta **36**, 827 (1963). — LANDÉ, A.: Phys. Rev. **108**, 891 (1957). — SCHILPP, P. (ed.): Albert Einstein: Philosopher-Scientist. Evanston: Library of Living Philosophers 1949. — SHIMONY, A.: Am. J. Phys. **31**, 755 (1963). — TISZA, L.: Revs. Mod. Phys. **35**, 151 (1963). — WIGNER, E. P.: Am. J. Phys. **31**, 6 (1963). An excellent, fully documented historical treatment of views here discussed is found in MAX JAMMER, The conceptual development of the quantum theory. New York: McGraw-Hill Book Co. 1966. — The contemporary scene is well surveyed by BERNARD D'ESPAGNAT, Conceptions de la Physique Contemporaine. Paris: Herman 1965.

[2] HEISENBERG, W.: Physics and philosophy. Harper 1958.

[3] MARGENAU, H.: The nature of physical reality, p. 171 et seq. New York: McGraw-Hill Book Co. 1950.

[4] — Phil. Sci. **30**, 1, 138 (1963).

[5] — The nature of physical reality. Ref. 3, chap. 13. Further discussion of the objectivity problem may be found. In: H. MARGENAU and J. PARK: Objectivity in quantum mechanics. In: M. BUNGE, ed.: Delaware Seminar in the Foundations of Physics. New York: Springer 1967.

[6] JEFFREYS, H.: Scientific inference. New York: McMillan 1931.

[7] MARGENAU, H.: Ann. of Phys. **23**, 469 (1963).

[8] —, and R. N. HILL: Progr. Theoret. Phys. (Kyoto) **26**, 722 (1961).

[9] WIGNER, E.: Phys. Rev. **40**, 749 (1932).

[10] COHEN, L.: J. Math. Phys. **7**, 781 (1966).

[11] — Thesis Yale University 1966, University Microfilm, Ann Arbor, Michigan.

[12] GROENEWOLD, H. J.: Physica **12**, 405 (1946).

Chapter 5

Hidden Parameters Associated with Possible Internal Motions of Elementary Particles

Jean-Pierre Vigier

Institut Henri-Poincaré, Paris, France

Most of the past discussion on hidden parameters has concentrated on the possibility of introducing dispersionless variables into quantum theories. Here I would like to discuss another possible type of hidden parameters, namely those which could be associated with non hermitian unitary transformations. These appear naturally if:

a) the internal symmetries of elementary particles are really connected with internal dynamical motions; and

b) if these symmetries turn out to be non-compact, as various authors have recently suggested.

As an illustration of such a possibility I will discuss some models all directly or indirectly connected with the question of the relation of the external invariance groups (such as P or $SO(4,1)$) and internal symmetries (describing strong interactions). Elementary particles have been found to be described by various quantum numbers, some of which (J^P and M^2 that is spin and mass) are related with the two invariants of the external group of motion (P or $SO(4,1)$) while the others — T (isobaric spin), S (strangeness) and B (baryon number) — are associated with an internal group (such as $SU(3)$) and are usually denied any dynamical signification. Usually one assumes the total symmetry to contain $P \otimes SU(3)$ multiplets and the observed situation has to be interpreted as a breaking of the internal symmetry. Another point of view is possible however: namely to unify external and internal Lie Algebras (so that the mass operator can split $SU(3)$) within a single Lie Algebra. In the last example discussed here, we shall see that one can unify $SO(4,1)$ with a non compact internal symetry group $SU(2,1)$ within the usual conform group $SU(2,2)$, and obtain a mass splitting without any symmetry breaking [1].

This of course would introduce new features into the interpretation of wave mechanics. Not only would we introduce new variables associated with the new quantum numbers needed to describe elementary particles, but we would have to modify deeply the present measurement theory within distances $\leq 10^{-14}$ cm assumed to limit the internal space-time of elementary particles.

1. Groups of Motions on Riemannian Manifolds

I shall now develop some possible applications of these ideas. However, before I do this I want to recall that the general mathematical theory of groups of motions on Riemannian manifolds has been extensively studied by mathematicians (such as Fubini [2], E. Cartan [3], Vranceanu [4], Egorov [5], Ishihara [6], Lichnerowicz and K. Yano [7]) so that the physicist's work in this field is mainly a problem of physical interpretation of known mathematical results.

For example on a V_4 depending on the choice of the stability subgroup we get[1]:

1°) 10 parameter groups of motions G_{10} (that is G_r with order $r = 10$), namely $SO(5)$, $SO(4,1)$, $SO(3,2)$, $(R_4 \times T_4 = P)$; the corresponding Riemannian spaces having a constant curvature. This results immediately from the fact that in a V_n the maximum order of a possible group of motion G_r is $r = n(n+1)/2$, which is only reached when V_n is of constant curvature.

2°) Since no G_r with $r = 9$ exists, the next set is a set of G_8's namely SU_3, $SU(2,1)$, $T_{2c} \times U_2$ where T_{2c} denotes translations in a 2 dimensional complex space.

Both sets 1°) and 2°) correspond to Einstein spaces $(R_{ij} = \Lambda g_{ij})$ with and without constant curvature.

3°) The next set is a set of G_7's namely:

$$T_1 \times R_4, \quad T_1 \times SO(3,1), \quad T_1 \times [T_3 \times R_3], \quad T_{2c} \times SU_2.$$

4°) For G_6's we get:

$$R_3 \otimes R_3; \quad SO(2,1) \otimes SO(2,1); \quad R_3 \otimes [T_2 \times R_2]$$

$$SO(2,1) \times [T_2 \times R_2] : [T_2 \times R_2] \otimes [T_2 \times R_2].$$

4°) For G_5 we have only:

$$T_1 \otimes T_1 \otimes R_3.$$

And we have other groups G_r with $r < 5$.

Of course similar classifications can be performed for $n > 4$. For example, for $n = 5$ we discover among others, in categories 1°) and 2°), the groups $G_{15} = SU(2,2) \simeq SO(4,2)$ and the Weyl groups $G_{11} = P \times A_0$

---

[1] $\times$ denotes semi direct product and $\otimes$ the direct product.

and $G_{11} = SO(4,1) \times A_0$ (which can be used in Kaluza-Klein [8] or Jordan-Thiry [9] types of theories) the corresponding manifolds beeing also Einstein-spaces with or without constant curvature.

I shall now summarize four concrete examples of application of the preceding considerations to Elementary Particle Theory.

2. Interactions and Manifolds

If we remark that the introduction of gravitational masses in an empty space gives rise to gravitational interactions which change the geometry of the space itself, we may expect that the strong interaction of elementary particles will also change the geometry of an interaction region.

In order to apply this idea to elementary particle theory Raçzka [10] chooses for this interaction region a V_4 manifold invariant under the group $G_4 = T_{x_0} \otimes T' \otimes R'_3$ in order to have the usual time-like displacement, the V_4 beeing compact with repect to three space variables. In that case V_4 is homeomorphic to the direct product of a straight line (time) R one dimensional torus S_1 and two dimensional sphere S_2. The ds^2 of this space is:

$$ds^2 = -(dx_0)^2 + R_T\,d\alpha^2 - R_s^2(d\beta^2 + \sin^2\beta\,d\gamma^2).$$

On this V_4 the Laplace-Beltrami equation:

$$\left[\frac{1}{\sqrt{|g|}}\,\partial_\nu\,g^{\mu\nu}\,\sqrt{|g|}\,\partial_\nu\right]\Psi = x^2\,\Psi$$

where $g_{\mu\nu}$, g, and $\partial_\nu = \partial/\partial x_\nu$, represent the usual symbols, becomes:

$$\left[-\frac{\partial}{\partial X^2} + \frac{1}{R_T^2}\,\frac{\partial^2}{\partial\alpha^2} - \frac{1}{R_s^2\sin\beta}\left(\frac{\partial}{\partial\beta}\sin\beta\,\frac{\partial}{\partial\beta} + \frac{1}{\sin\beta}\,\frac{\partial^2}{\partial\gamma^2}\right)\right]\varphi = x^2\,\varphi.$$

Its solutions can be developed on a set of orthonormal functions:

$$\Psi = e^{iEx_0}\,e^{iY\alpha}\,Y_{T,T_3}(\beta,\gamma)$$

subject to the condition that E (energy), Y (hypercharge) and T (isobaric spin) satisfy the relation:

$$E^2 = Y^2 - \frac{1}{R_s^2}\left[T(T+1) - \left(\frac{R_S}{R_T}\right)^2 Y^2\right]$$

which is just Okubo's mass formula for bosons if we take $R_s/R_T = \frac{1}{2}$ and $R_s \simeq 1/2{,}5\,m_\pi$. The interest of this very simple calculation is that we obtain a mass formula without any breaking of symmetry. Its main defect is of course that we only obtain integer values of T on S_2. However one could remedy this defect as we shall discuss later by passing to a V_5 and a G_5 such that $G_7 = T_{x_0} \otimes T'_1 \otimes SU(2)$ with $V_4 = R \otimes S_1 \otimes S_3$.

3. Quantum Numbers and Curved Space

A second problem can be treated with the help of curved Riemannian manifolds. It is known that, at least for bosons, there exist particles with identical J^P, T, T_3 and Y but with different masses. This suggests that one should introduce something like a principal quantum number in order to remove the degeneracy [11]. We shall now see that such a quantum number can arise naturally from the consideration of a curved space, the group of motions of which can be considered as a "degeneracy-removing group".

It is well known that the maximal order of the groups of motions of a Riemannian n-manifold V_n is $\frac{1}{2} n(n+1)$, the maximal order being obtained for constant-curvature spaces (spheres, pseudospheres or flat spaces). In order to have a coherent mass spectrum, we shall consider a V_4 as $S^3 \times R$ (the so-called Einstein model), which can be embedded isometrically in R_5 as a hyperspherical cylinder, S^3 being the hypersphere $z_1^2 + z_2^2 + z_3^2 + z_4^2 = r^2$ in $R^4(z_1, \ldots, z_4)$. We take on V_4 the metric: $ds^2 = dt^2 - d\sigma^2$ where $d\sigma^2$ is the natural metric of S^3 in the four-dimensional Euclidean space E_4. Naturally the group of motions of our V_4 is $O_4 \times R$.

One deduces therefrom the Laplace-Beltrami operator, $\square = \partial^2/\partial t^2 - (1/r'^2) L$, where L is defined by $r'^2 \Delta = (1/r')(\partial/\partial r')(r'^3(\partial/\partial r')) + L$ (Δ being the usual Laplacian in E_4).

We denote by P a nth order homogeneous harmonic ($\Delta P = 0$) polynomial, and define the hyperspherical function of degree n, Y_n, by $P(z_1, \ldots, z_4) = r^n Y_n$. Then Y_n satisfies the partial differential equation:

$$L Y_n + n(n+2) Y_n = 0.$$

In addition one can show that:

$$\int_{S_3} Y_n Y_n' \, d\Omega = 0 \quad \text{for } n \neq n'.$$

For a fixed n, we have:

$$\binom{n+3}{3} - \binom{n+1}{3} = (n+1)^2$$

linearly independent Y_n's.

These Y_n in polar co-ordinates $(r, \vartheta_1, \vartheta_2, \varphi)$ of E_4 are polynomials in sines and cosines of ϑ_1, ϑ_2, φ.

If we now look for the levels $-m_0^2$ in the Hilbert space generated by functions $\Psi = e^{imt} Y_n$ with the topology of $L^2(S^3 \times R)$, we get $m_0^2 + (1/r^2) n(n+2) = m^2$. These are, therefore, the eigenvalues of our static solution. Mathematically we may consider the $V_4 = H^3 \times R$, where H^3 is the single sheet hyperboloid $z_1^2 + z_2^2 + z_3^2 - z_4^2 = r^2$ (embedded in the pseudo-Euclidean space E_1 isometrically), and we can make the same argument (substituting φ into $i\varphi$). In this case the group of motions will be $O(3,1) \times R$. In order to have functions which vanish when $\varphi \to +\infty$ (i. e. at infinity), we the $\frac{1}{2} n(n+1)$ functions (for fixed n) $e^{-k\varphi p} P^{(n,k)} (0 < k \leq n)$, and then consider the Hilbert space generated by the functions:

$$\Psi = e^{imt} \cdot e^{-k\varphi p} P^{(n,k)} (x_1, x_2).$$

Continuing in the same manner as before, we obtain for the eigenvalues $-m_0^2$ of the Laplace-Beltrami operator the same formula $m^2 = m_0^2 + (1/r^2) n(n+2)$.

Notice that in both cases, in the limit of flat space ($r \to \infty$), we get the basic level $m_0^2 = m^2$. The introduction of curvature in space removes the "degeneracy" by giving rise to supplementary levels.

Notice also that in both cases for a given n there is just one function among the $(n \not\equiv 1)^2$ [or the $\frac{1}{2} n(n+1)$ in the second case], which is only a function of the distance in the curved space. (For instance in the first case this is the function $(\sin(n+1)\varrho/r)/((n+1)\sin\varrho/r)$, where ϱ is the distance on the three-dimensional sphere of radius r.)

It was indicated empirically that a mass formula containing a phenomenological term of an integer multiple of the pion mass gives a good description of the mass spectrum of strongly interacting particles.

If we take the particular choice of parameters, so that for $n=0$, $m = \sqrt{\langle m_\pi^2 \rangle}$, and that $r = 1 \sqrt{\langle m_\pi^2 \rangle}$ we obtain

$$m = \sqrt{\langle m_\pi^2 \rangle} (n+1) \qquad (n = 0, 1, 2 \ldots).$$

4. Particles as Warps in a V_5

In the third example we want to introduce a new physical idea. As we know, interaction forces in Nature can be roughly divided into long range (gravitation and electromagnetism) and short range (strong). The former can be represented as isometric groups of motions on a weakly curved V_5. It is thus very tempting to try also to geometrize the latter and to represent particles as small strongly curved manifolds in an internal five dimensional space time. We thus start from the assumptions:

(*a*) That the external and internal motions of elementary particles are isometric groups of motions on *different* external five dimensional curved Riemannian manifolds (Lichnerowicz's domains of isometry).

(*b*) That these domains are connected through an isometry boundary (time-like hypertube) built with trajectories of the internal motions and geodetics of the external domain. On this boundary we unify, in the sense of Flato, Vranceanu [4] and Egorov [5], the external $P \times U(1)$

group with the internal $(SU(2,1))$ motion groups (also considered here as isometry groups) within the global symmetry group $SU(2,2)$ (6).

This introduces EINSTEIN's general relativity ideas into elementary particle theory. Particles are now considered as extended regions with very strong internal curvature; and we shall now show how to continue (in LICHNEROWICZ's sense [8]) the internal manifold through the boundary into a neighbourhood belonging to external (event) space-time. This connects internal and external curved space-times according to the requirements of KOMAR [12], TREAT [13] and LOOS [14].

We go further along this line in the following way.

(a) Let g be the group of stability of a four dimensional homogeneous Riemanian space $V_4 = G/g$ admitting the motion group G. If g is isomorphic to the real representation of $U(2)$ then V_4 is a Kählerian space with constant holomorphic sectional curvature. If V_4 is connected and simply connected then V_4 is homeomorphic to $P(2,C)$ (two dimensional complex space) and G is a G_8 locally ilomorphic to $SU(2,1)$. The ds^2 of this V_4 has been calculated by EGOROV [13]. We get:

$$ds^2 = (\exp -2A)\,[(y_2\,dy_1 - y_1\,dy_2)\,(\bar{y}_2\,d\bar{y}_1 - \bar{y}_1\,d\bar{y}_2) \\ - dy_1\,d\bar{y}_2 + d\bar{y}_1\,dy_2)] \tag{1}$$

with $\exp A = (y_1\,\bar{y}_1 - y_2\,\bar{y}_2 - 1)$, the bar denoting complex conjugation. Our V_4 is parametrized with four real variable x_i ($i = 1, 2, 3, 5$) with $y_1 = x_1 + ix_2$, $y_2 = x_3 + ix_5$ and has normal elliptic signature. This V_4 we now consider as the space-like section of our internal V_5 which we take with a $ds'^2 = dt^2 - ds^2$: that is, with a five dimensional normal hyperbolic signature.

This V_4 section is an Einstein space satisfying $R_{ij} = \Lambda\,g_{ij}$, where Λ represents a very strong internal curvature. The choice of $SU(2,1)$ instead of $SU(3)$ first suggested by FLATO and all [6] has been shown (18) to recover all known physical results obtained through $SU(3)$.

(b) We now define the boundary as a hypertube $V_3 \times R$. V_3 limits the V_4 of (1) and R is a straight line corresponding to time. To build it with trajectories of internal motions it is sufficient that the G of this V_3 correspond to $g = U(2)$. This means our G is a G_4. As shown by VRANCEANU [4] a V_3 which admits a G_4 takes the form $R \times S_2$, where R represents a flat curve and S_2 a two dimensional surface with constant curvature. This means our boundary can be taken with:

$$ds''^2 = R^2\,d\alpha^2 + R^2(d\beta^2 + \cos^2\beta\,d\gamma^2 + \sin^2\beta\,d\varrho^2) - dt^2 \tag{2}$$

R can represent the particle radius $\ll 10^{-13}$ cm.

Our boundary is thus locally invariant at each point, also under the maximal compact subgroup $SU(2) \times U(1)$ of $SU(2,1)$, the factors corresponding respectively to T-spin and $Y/2$. The baryon number corresponds at each point of our internal manifold to the real representation of the center of $SU(2,1)$.

5. Fitting the Internal and the External Space-Times

The last example is an attempt to represent geometrically the unification procedure. The external dynamical group $SO(4,1)$ and internal symmetry group $SU(2,1)$ are both considered to be isometry groups of motions on a V_5 which represents an internal space-time (denoted Σ_5) we "fit" on an external space time V_5 invariant under the Weyl group $SO(4,1) \times U(1)$, which corresponds to the combined effects of gravitation and electromagnetism.

If we recall a well known result of Kobayashi, we see that the 5-dimensional hyperbolic sphere Σ_5 can carry isometric transformations which yield the Lie Algebra of the locally isomorphic groups $SO(4,2) \simeq SU(2,2)$. That is a G_{15}. This Σ_5 can be considered as a surface embedded into a six dimensional space $E(4, 2)$. Indeed we can parametrize it with

$$x_1 = \text{ch } \vartheta \, \cos \xi \, \sin \varPhi_2 \qquad x_2 = \text{ch } \vartheta \, \cos \xi \, \cos \varPhi_2$$

$$x_3 = \text{ch } \vartheta \, \sin \xi \, \cos \varPhi_3 \qquad x_5 = \text{ch } \vartheta \, \sin \xi \, \sin \varPhi_3$$

$$x_0 = \text{sh } \vartheta \, \cos \varPhi_1 \qquad x_6 = \text{sh } \vartheta \, \sin \varPhi_1$$

with

$$-\infty \leq \vartheta < \infty, \quad 0 \leq \xi < \frac{\pi}{2}, \quad 0 \leq \varPhi_i \leq 2\pi, \quad (i \neq 1, 2, 3).$$

there surface beeing defined by the relation:

$$x_1^2 + x_2^2 + x_3^2 - x_0^2 + x_5^2 - x_6^2 = 1$$

with

$$ds^2 = -d\vartheta^2 - \text{sh}^2 \vartheta \, d\varPhi_1^2 + \text{ch}^2 \vartheta \, (d\xi^2 + \cos^2 \xi \, d\varPhi_2^2 + \sin^2 \xi \, d\varPhi_3^2).$$

We can then calculate the 15 generators of our G_{15} namely $H_{\alpha\beta} = x_\alpha \, \partial_\beta - x_\beta \, \partial_\alpha \, \alpha, \beta = 1 \ldots 6)$ and construct the Weyl basis of our algebra. The construction of the generators of $SO(4,1)$ and $SU(2,1)$ is then straightforward and one can construct the solutions of the Laplace-Beltrami operator (or of its linearized form $\Gamma_\alpha \Gamma_\beta M^{\alpha\beta} \varPsi = x^2 \varPsi$ with $[\Gamma_\alpha, \Gamma_\beta]_+ = 2g_{\alpha\beta}$). Evidently the mass operator splits the multiplets of $SU(2,1)$ and one recovers an empirical mass formula already presented elsewhere [15].

The fitting of external and internal space time can be performed on a de Sitter space as anticipated by Barut and Bohm. Furthermore on the corresponding isometry boundary the $g_{\mu\nu}$ of the constant curvature internal space-time (satisfying $R_{\alpha\beta} = \Lambda g_{\alpha\beta}$) are homothetic to those of the external space-time (also satisfying $R_{\alpha\beta} = \lambda g_{\alpha\beta}$) with non constant curvature. The particle is thus comparable to a potential well containing discrete levels of energy (mass) corresponding to different types of elementary particles.

This model also modifies the question of the behavior of particles which no longer appear as resulting from nonlinear terms in the wave equations. The problem of their possible spreading vanishes: particles are held together (similarly to geons) by their own strong curvature. The preceding continuation theory explains why they are also "guided" by external fields. The observed baryon spectrum results in this model of quantized internal behaviour corresponding to irreducible finite dimensional representations of our groups of motions. Mass itself results of internal subquantum motion. Observed baryons just correspond to internal quantized behavior which yield a possible physical description of the "clocks" attached to material particles since the foundation of wave mechanics. In this light it is remarkable that the known bosons can be considered as non massless quanta emitted when a baryon undergoes quantum jumps from one quantum state to another.

References

[1] FLATO, M., D. STERNHEIMER, and J.-P. VIGIER: Compt. rend. **260**, 3869 (1965).
[2] FUBINI, G.: Ann. Mat. (3) **8**, 39 (1909).
[3] CARTAN, E.: Bull. soc. math. France **54**, 14 (1921).
[4] VRANCEANU, G.: Studii s. Certari Math. **4**, 121 (1953).
[5] EGOROV, I. P.: Dokl. Akad. Nauk U.S.S.R. **9**, 103 (1955).
[6] ISHIHARA, S.: J. Math. Soc. Japan **7**, 345 (1955).
[7] YANO, K.: Theory of lie Derivatives. Amsterdam: North-Holland 1957.
[8, 9] LICHNEROWICZ, A.: Théories relativistes de la gravitation et de l'électromagnétisme. Paris: Masson & Cie 1955.
[10] RACZKA, R.: Int. Cent. Theor. Phys. Report No. IC/65/32 to be published.
[11] FLATO, M., G. WATAGHIN, D. STERNHEIMER, and J.-P. VIGIER: Nuovo Cimento **42**, 431 (1966).
[12] KOMAR, A.: Phys. Rev. Letters **13**, 220 (1964).
[13] TREAT, R. P.: Phys. Rev. Letters **12**, 407 (1964).
[14] LOOS, H. G.: Ann. Phys. **36**, 486 (1966).
[15] FLATO, M., and J. STERNHEIMER: Compt. rend. **259**, 3455 (1964).

Chapter 6

An Axiomatic Foundation of Quantum Mechanics on a Nonsubjective Basis

G. Ludwig

Institut für theoretische Physik der Universität Marburg, Germany

In the context of quantum mechanics we use to speak of observables, properties, states, probability, measuring-process, or of microscopic objects such as electrons, protons, atoms, etc. Nobody can formulate the meaning of these words with *absolute* precision and only the physicist has some sort of "feeling" for them.

On the other hand, the Hilbert space as a beautiful mathematical scheme for quantum mechanics is only the final result of a long historical development. Therefore, it seems necessary to look for fundamental physical structures which are "strong" enough to deduce from them the Hilbert space as a mathematical model. In the following I will try to describe just a start for a new foundation of a physical theory such as quantum mechanics.

The concepts of microscopic objects like electrons, protons or atoms, have evolved in a long historical process of thinking about numerous physical problems. Here we shall try to understand from a *systematic* point of view why it is possible to speak of physical objects like protons, atoms etc. Let us first discuss the behavior of macroscopic objects, assuming it is given in an objective way.

We do not attempt to analyze the philosophical question of speaking of "the cup on the table" and that it is not necessary to use "I have seen a cup on the table". *Every experimental test is based on such objective facts.* The aggregate of these objective facts are called briefly the macroscopic region.

What we are interested in in physics is the influence of one physical object on another. Let us take a piece of uranium and a cloud-chamber for example. Then the piece of uranium will influence the cloud-chamber. Usually we speak of the uranium as emitting particles that produce the tracks in the cloud-chamber, but here we shall avoid this formulation and concentrate only on the macroscopic objects. Yet we will not discuss just individual objects like that particular piece of uranium, but shall take into account the fundamental procedure in physics, namely to repeat many times "the same experiment". So if we speak of a piece of ten grams of uranium, we do not refer to any specified piece but to the class of such pieces.

A class of macroscopic objects is defined by a physical or technical process which produces the objects. So the class is the output of well-defined physical procedures. But this does not mean that all macroscopic objects given by a procedure are "equal". It could be possible that one procedure makes a sub-class of another procedure.

To develop a mathematical picture we consider two sets. The set K shall be the set of classes of macroscopic objects which influence the other macroscopic objects. The second set L shall be the set of all effects that can be produced by the objects of K on other macroscopic objects. The case in which several effects can be produced on the same macroscopic object will be very important in our analysis and will be treated later on.

Typical effects are the signals of a counter or the water-droplets of a cloud-chamber track. Taking an element V of K, we can observe the relative frequency of an effect F produced by this macroscopic object V. This relative frequency will be called the *probability* for V to produce F. Since our time is too limited for a detailed discussion of physical probability this short illustration may be sufficient.

The preceding statements are formulated in the first axiom:

Axiom I a': There is a function $\mu(V, F)$ on the product-set $K \times L$, such that the values of μ are rational numbers: $0 \leq \mu \leq 1$.

With the help of this function μ, the sets K and L can be subdivided into equivalence classes:

V_1 and V_2 shall belong to the same class if $\mu(V_1, F) = \mu(V_2, F)$ for all $F \in L$. F_1 and F_2 shall be elements of the same class if $\mu(V, F_1) = \mu(V, F_2)$ for all $V \in K$.

These equivalence classes shall form new sets K and L respectively, and the elements of K will be called *ensembles*, those of L *effects*. For the moment the term 'ensemble' ist just a word; the reason for choosing it will become clear in the following.

Now instead of axiom I a', we can take axiom 1 a:

Axiom 1 a: There is a function $\mu(V, F)$ on the product-set $K \times L$, the values of μ being rational numbers $0 \leq \mu \leq 1$.

If $\mu(V_1, F) = \mu(V_2, F)$ for all $F \in L$, then $V_1 = V_2$.

If $\mu(V, F_1) = \mu(V, F_2)$ for all $V \in K$, then $F_1 = F_2$.

The division of K and L into equivalence classes is at first merely formal. It becomes a statement about physical structure through the fact that all the following axioms are related only to the set $K \times L$, the set $\boldsymbol{K} \times \boldsymbol{L}$ being no longer involved: The fact that only the elements V of K and not the $\boldsymbol{V}$ of $\boldsymbol{K}$ enter all the following relations is briefly expressed by talking of 'objects' emitted by $\boldsymbol{V}$ objects, which produce the effects. It is not the specific apparatus $\boldsymbol{V}$ that goes into the physical structure but only the equivalence class V. For instance, we speak of light of a sort V irrespective of the special producing source $\boldsymbol{V}$.

The further possibility of combining different V_s' to a wider class of objects, called 'objects of the same kind', is only the consequence of additional physical structures which will be discussed later on. Also historically, such unification sometimes occurs only afterwards, for example through the discovery of light and electromagnetic waves being 'of the same kind'.

Since we do not consider a single experiment with an apparatus $\boldsymbol{V}$ from class V but a numerous repetition of experiments for statistical purposes, V will not be called merely an object but 'an ensemble of objects' or briefly an 'ensemble'.

The circumstance that all following axioms also contain only the equivalence classes F, not the concrete effects $\boldsymbol{F}$, expresses a sort of objectivity of the effects. This shall mean that all effects $\boldsymbol{F}$ of the same class F are caused by some objective structure of the objects. Nevertheless, we have good reason not to call the classes F 'properties', and will comment on it subsequently. Since for the following axioms only the classes F are relevant, we will not coin a new word; but the elements F from L shall be called effects also. As it is possible to count the number of experiments we set up

Axiom 1 *b*: There is an F with

$$\mu(V, F) = 1 \qquad \text{for every } V.$$

Merely for reasons of mathematical aesthetics we add a purely formal axiom:

Axiom 1c: There is an F (Def. 0) with

$$\mu(V, 0) = 0 \qquad \text{for every } V.$$

Now we can combine a number N_1 of experiments of species V_1 and N_2 experiments of species V_2 to form $N = N_1 + N_2$ experiments, which motivates the following

Definition: For every pair V_1, V_2 in K and every rational number λ with $0 \leq \lambda \leq 1$ we define a V in K by $\mu(V, F) = \lambda \cdot \mu(V_1, F) + (1 - \lambda) \cdot \mu(V_2, F)$ for all F.

The element V is called a 'mixture' of V_1 and V_2.

Due to lack of space the subsequent axiom is not quite precise mathematically. The exact formulation can be seen in [1]. Let us mention only briefly that in analogy to the process of 'completion' for the rational numbers giving the real numbers, one can 'complete' K and L mathematically. An order relation is established in L through the definition: $F_1 \leq F_2$ if $\mu(V, F_1) \leq \mu(V, F_2)$ for all V in K. Be $K_-(F)$ the subset of all V in K with $\mu(V, F) = 0$. Now we use the foregoing definitions to write down the very important axiom of 'sensitivity increase':

Axiom 2: For every F_1 and F_2 in L there exists an F_3 in L such that $F_1 \leq F_3, F_2 \leq F_3$ and $K_-(F_3) \supseteq K_-(F_1) \cap K_-(F_2)$. For every boundary element V of K there is a $F \neq 0$ with $\mu(V_1 F) = 0$.

This axiom states that, provided a supplementary condition is satisfied for effects F — namely not to be produced by certain objects — then it shall be possible to increase the sensitivity of effects in such a way that for every two of them F_1, F_2 there exists an F_3 of greater sensitivity.

Let $L_-(k)$ be the set of all F in L with $\mu(V, F) = 0$ for all V from $k \subseteq K$. Axiom 3 then implies the existence of a maximal effect E in $L_-(k)$ with $F \leq E$ for all F in $L_-(k)$. These maximal effects shall be called 'decision effects' (the term 'propositions' is often used by others). The word 'property' will be reserved for a later use.

It follows without any further assumption that the set of decision effects with the order relation defined in L forms a complete lattice G and that from $\mu(V, E_1) = 0$, $\mu(V, E_2) = 0$ follows $\mu(V, E_1 \cup E_2) = 0$ for every pair E_1, E_2 of G. Axiom 3 explains the curious circumstance that the decision effects can be considered as elements of a lattice G.

Introducing the concept of commensurability we obtain another important structure. This is suggested by the following technical possibility: the same object shall produce two effects F_1, F_2 on one apparatus. Then we can construct a 'sum effect' $F_1 + F_2$ ($F_1 + F_2$ is there if F_1 or F_2 are present but not both) and a 'product effect' $F_1 \cdot F_2$ ($F_1 \cdot F_2$ being there if both, F_1 and F_2 are present). With these definitions at hand we can form a Boolean ring from F_1 and F_2. A Boolean ring is defined generally to be a ring with the additional property $a \cdot a = a$ for every element a.

Corresponding to the technical possibilities outlined above and the immediate implications for $\mu(V, F)$ we shall define: a set A of decision effects E is called commensurable if there is a Boolean ring R with $A \subseteq R \subseteq G$ and

$$\mu(V, E_1) + \mu(V, E_2) = 2\mu(V, E_1 \cdot E_2) + \mu(V, E_1 + E_2)$$

for every pair of elements E_1, E_2 in R and all V in K. We may draw attention to the fact that in the first instance the operations '$+$' and '$\cdot$' are not related to the lattice operations in G. The definition given above requires only to find two operations '$+$' and '$\cdot$', such that R becomes a Boolean ring and μ satisfies the corresponding relation. When it is possible mathematically to imbed A into R the decision effects from A are called *commensurable*.

This formal definition becomes a statement about the physical structure through associating the mathematical concept "commensurable" with the actual possibility of constructing an apparatus on which the E in A can be produced by one object. So a mathematically possible construction is put in correspondence with a physically possible construction.

The following theorem can be proved [1]:

Two decision effects E_1, E_2 with $E_1 \leqq E_2$ are commensurable.

This theorem has the following consequences: the lattice G is ortho-complementary and orthomodular, the measures $\mu(V, E)$ on G are ortho-additive, i. e.

$$\mu\left(V, \bigcup_i E_i\right) = \sum_i \mu(V, E_i)$$

provided the E_i are pairwise orthogonal. Summarizing our exposition we may remark that the axioms 1 to 2 involve all the essential properties of a physical probability field.

Let me finally comment on some physical concepts, still to be deduced from the foregoing considerations.

As *centre Z* of the lattice G we designate the set of all elements of G which are commensurable with every element of G. The elements of Z shall be called 'properties', since they can be associated with the objects without reference to the 'measuring apparatus'.

All objects with the same 'properties' will be said to be 'of the same kind'. In this sense, light and radiowaves are of the same kind, but electrons and protons of different ones.

One often combines several such kinds to form a class, if certain physical transformations convert one kind into another. Then the classes of objects are determined through the transformation group under consideration. So a further physical structure emerges, characterized just by the group of transformations, but here we shall not enter into this question. Only different kinds of objects are discussed in the following. The existence of a centre Z corresponds to the possibility of representing all ensembles V as mixtures of ensembles V_α, the V_α having no dispersion with respect to the 'properties' from Z.

Two examples may serve as illustration of our general reflections.

1. *Classical point-mechanics.* The elements V of K are realized here by all measurable functions $V(p)$ in phase space (Γ-space) with $V(p) \geqq 0$ and $\int V(p)\, dm(p) = 1$, m being the Lebesgue measure. The effects $F \in L$

are given by all measurable functions $F(p)$ in Γ-space with $0 \leq F(p) \leq 1$. Our expression for $\mu(V, F)$ is $\int V(p) F(p) \, dm(p)$. The decision effects E are characterized by all such $E(p)$ that $E(p) = 1$ or 0, thereby corresponding to the measurable subsets of Γ. The centre Z of G is the whole G.

2. *Quantum mechanics of objects of the same species.* The elements V of K are now all Hermitean operators $V \geq 0$ with $\mathrm{Tr}(V) = 1$, the effects $F \in L$ all Hermitean operators F with $0 \leq F \leq 1$ and the function $\mu(V, F) = \mathrm{Tr}(VF)$. Decision effects $E \in G$ are given by projection operators E and hence can be characterized by subspaces of the Hilbert space, such that G may be identified with the lattice of subspaces. The centre Z of G just contains $E = 0$ and $E = 1$ or the zero-space and the whole Hilbert space.

In the usual representations of quantum mechanics one so far introduces only the lattice G of projection operators respectively the observables A, A being an Hermitean operator that is not restricted by a supplementary condition (in contrast to the F). Quite deliberately we did not talk of observables up to now. Let me emphasize here that the effects F should be confused neither with the observables A nor with the ensembles V, although the F as well as the A and the V are represented by Hermitean operators. Also in the general theory as formulated by the foregoing axioms we can introduce observables A besides the $F \in L$ and $E \in G$, A being nothing but the integration of a set $g \subseteq G$ of commensurable decision effects making use of a scale α. α is not given by an objective physical structure but is only particularly suitable for theoretical description so that we can write:

$$A = \int \alpha \, dE_\alpha$$

in the sense that the expectation value of A becomes

$$\mu(V, A) = \int \alpha \, d\mu(V, E_\alpha).$$

In establishing quantum mechanics, the use of all effects F instead of decision effects E only is profitable also for the discussion of the measuring process. Without going into detail, let me outline the question very briefly.

We began our axiomatic procedure with considering effects produced on macroscopic objects without asking for the detailed mechanism of the production. The problem of "measurement" is now the following. Given a specific theory (for example quantum mechanics including the real interactions), one has to treat as a new object of the theory the combination of microobject plus macroobject, thereby facing the necessity of considering ensembles of new objects (composed of macro- and microobjects). The effects, originally fixed to the macroobjects, must now be interpreted as possible actions on the environment of the new object.

A more elaborate study of this question now shows that the actual course of an experiment in general does not proceed through production of a decision effect, but the effect fixed on the macroobjects can be represented by Hermitean operators F with $0 \leq F \leq 1$ which refer only to the microobjects.

The result of an action of a microobject on a macroobject can always be represented exclusively in the Hilbert space of the microobject, though in general only by a Hermitean operator F with $0 \leq F \leq 1$. Representation through a projection operator is possible in limiting cases only, being attained very seldom in reality.

To predict, from the factual interaction possibilities the effects F actually realizable, would be the problem of a theory of measurement still to be solved. The general theory. as formulated by the given four axioms, seems to pose a very interesting problem also for mathematicians namely to find all isomorphic models satisfying the axioms.

References

[1] Ludwig, G.: Versuch einer axiomatischen Grundlegung der Quantenmechanik und allgemeinerer physikalischer Theorien. Teil I: Z. Physik **181**, 233 (1964); — Teil II: Preprint Marburg, June 1966, to be published in Communications in Math. Physics.

Chapter 7

A Ghost-Free Axiomatization of Quantum Mechanics[1]

Mario Bunge

Physikalisches Institut der Universität Freiburg i. Br., Germany[2]

In this paper elementary quantum mechanics (henceforth QM) is given a formulation characterized by two traits that do not seem to have occurred jointly before, namely logical order and a strictly physical interpretation. In other words, the formalism of QM is organized axiomatically and is assigned a content free from nonphysical (particularly psychological) associations. The details are given elsewhere [*1*].

1. To Look or not to Look: this is not the Question

The subjectivist (operationalist, positivist, phenomenalist) wishes to exorcize HAMLET's ghosts by substituting "To look or not to look" for "To be or not to be". This is what he does when he claims that the question of the real (autonomous) existence of atoms is meaningless or metaphysical, when he holds that the behavior of every atom — even the most forlorn atom in the center of Sirius — is determined by our measurement set-ups, and when he contends that the state of an atom will jump after the measurement interaction is over, just because the observer looks at the pointers.

In this way the subjectivist summons more ghosts than those which haunted HAMLET. Indeed, the claim that things acquire their properties just because we condescend to look at them is sheer anthropocentrism and, in order to be carried out consistently, it requires filling the whole cosmos with a staff of observers ever ready to take infinitely precise measurements of anything conceivable — just to keep the world going. And this is merely a modern version of animism.

The usual interpretation of QM — and also of relativistic theories — teems with the ghostly: the theoretician's demand that every symbol,

---

[1] This work has been supported by the Alexander von Humboldt-Stiftung.

[2] Presently at the Department of Philosophy, McGill University, Montreal, Canada.

even if it performs a purely computation function, be correlated to an experimental item; the observer's decision to look or not to look at the meter as decisive for the state of the object; the ideal measurement which nobody will ever perform; the "observables" which nobody can perceive; the interpretation of averages as expectation values and of scatters as uncertainties. All these are ghosts in the sense that they are not physical items. Most of them are in fact psychological ideas.

Surely every psychological item may constitute the object of psychological research. But this is irrelevant to physics for, by definition, physics studies physical systems, and physical systems are, by definition, self-existent, observer-free entities. Certainly not every physical object is given to us and none is grasped just as it is: we must gradually discover things and this requires making and correcting hypotheses about them. But the point is that, if they are not assumed to be out there, then they do not qualify as physical objects, i. e. as objects of study of physical science. Thus it is true that one will expect values near the center of a bell-shaped distribution to occur more frequently than others; but this does not render the average of a random variable a purely psychological attribute, a human expectation computed on the basis of subjective probabilities and nothing else.

2. Are the Ghosts Built into the Formalism of QM?

It is often claimed that a subjectivist epistemology is built into QM and that only an entirely new theory could restore the objectivist or realistic epistemology associated to classical physics. Moreover, it has sometimes been suggested that only a return to classical ideas could bring reality back to physics. This is demonstrably mistaken, for the question at hand is one of interpreting a given mathematical formalism, and formalisms are by definition neutral with respect to physical or psychological interpretation. Subjectivism can always be artificially injected into a physical theory and just as easily be eliminated from it.

Thus the objective statement

The properties x and y are related through $y = f(x)$ $\qquad\qquad \varphi_1$

can trivially be transformed into the psychological statement

Knowledge of x uniquely determines knowledge of y upon performing the computation indicated by "$y = f(x)$". $\qquad\qquad \psi_1$

Conversely, the psychological (or pragmatic) statement

The values that an observer can obtain upon measuring Q are the eigenvalues of the operator representing Q $\qquad\qquad \psi_2$

may be stripped of its pragmatic terms 'observer', 'obtain', and 'measuring', getting

The values of Q that a system may take on are the eigenvalues of the operator representative of Q. φ_2

On an anthropocentric philosophy the physical statements are either meaningless or abbreviations of their psychological partners. On a realistic philosophy, for the ψ_i to be true the φ_i must hold because the existence of the external world does not depend on its being perceived and computed. Surely, in order to *test* statements like φ_1 and φ_2 several additional pieces of knowledge and certain laboratory operations are needed. But it is one thing to test a statement and another to find out what it *means*, what it refers to, what it describes. It is one thing to make assumptions about physical systems and quite a different thing to make statements about the physicist's activities. Let the psychologist, the methodologist and eventually the physicist's wife advance judgments about him. Theoretical physics must be kept thoroughly physical, strictly ghost-free, or else its name must be changed into 'psychology'. For the concern of physics is not the observer and what he feels and thinks but the physical system. In particular, QM must be purged of its ghosts, which were artificially injected into it forty years ago by a positivist philosophy which hardly any living positivist holds nowadays.

Now, as everyone knows, ghosts can only be chased with light, the light of reason. And reason, though always critical, can be either chiefly destructive or mainly constructive. Having tried my hand at ghost chasing by kindling philosophical candles in the past [2], let me now present a constructive alternative which, though epistemologically classical (realistic), is thoroughly quantal in the sense that it keeps the basic stochastic character of QM and even proves that BORN's interpretation is unavoidable — unless the theory is enriched with nonrandom ("hidden") variables which, by not showing up in experiment, are as ghostly as the positivist intruders.

3. Background and Building Stones of a Realistic QM

Like every other physical theory, QM is built with the help of other theories, some of which are formal whereas others are physical. The formal presuppositions of QM are classical logic (the predicate calculus), analysis, and all the algebraic and topological presuppositions of the latter. The material background of QM consists of two distinct bunches of physical theories. One is classical electrodynamics, which of course should not be presupposed by quantum electrodynamics. The other material component of the background of QM is a collection of generic theories presupposed by several other specific theories as well, namely the theory of universal time [3], Euclidean physical geometry, the part-whole theory (which elucidates among others the concept $+$ of physical addition), and the physical models of probability theory, in particular the propensity interpretation [4], which renders probability theory applicable to single events of a kind.

The *primitive base* (aggregate of undefined concepts) of the theory shall be the 17-tuplet

$$B = \langle \Sigma, \overline{\Sigma}, E^3, T, \mathfrak{H}, \{Q\}, \{\mathcal{Q}\}, \{\varphi_k\}, \{c_k\}, H, \hbar, \mathcal{M}, V, A_0, A, c, \mathcal{E} \rangle.$$

Our postulates will glue these symbols together and will specify both their form (mathematical status) and their content (physical meaning). They will be presented in sections 4 and 7. Suffice it to say here that none of the above symbols stands for observers or for laboratory operations. In other words, the meaning of the basic symbols will not be specified in terms of observations and measurements. Firstly because empirical operations are not directly relevant to high-brow concepts such as the ones of hilbert space $\mathfrak{H}$ or hamiltonian H. Secondly because measurements do not determine meanings but they sample numerical values.

Basic QM refers to individual microsystems with quaint microproperties that cannot be caught by the eye. If observational concepts are wanted, then theories concerning perceptible things — in particular classical theories — can be resorted to. Such theories will occur in the test of QM but they cannot supply the meaning of QM, which is peculiar to it. Another way of making contact with the perceptible world is to apply basic QM to aggregates of myriads of atoms, in particular to those which make up real experimental arrangements — not the phony ones involved in gedankenexperiments or the ones dealt with by the half-born quantum theory of measurement. But the basic theory itself is free from observers and observations, however indispensable these are in the test stage. Should it be claimed [5] that the observer is the basic primitive of theoretical physics, then an axiom system specifying the characteristics of the observer and his operations should be supplied — but then this would not belong to theoretical physics and in any case no such system has been produced.

Our axiom system consists of two sets, one of comprehensive or model-independent assumptions, the other of specific assumptions concerning both the nature of microsystems and their interaction with their environment. This division is convenient because it shows how far one can go without making any special assumptions about the microsystems, much less about their milieu, whether natural or artificial. For example, one can in that way prove the stochastic character of QM and one can derive the general Heisenberg relations long before one has the slightest idea as to what a measuring instrument must look like if it is to take part in the test of some theorem of QM. To work.

4. Comprehensive postulates

A 1: Configuration space

1.1. E^3 is a 3-dimensional Euclidean space.

1.2. E^3 represents ordinary space.

A 2: Time

2.1. T is an interval on the real line R.

2.2. Every $t \in T$ represents an instant of time.

2.3. The relation $\leqq$ that orders T means "earlier than or simultaneous with".

A 3: Existence

3.1. Σ and $\bar{\Sigma}$ are nonempty denumerable sets.

3.2. Every $\sigma \in \Sigma$ is a microsystem (quanton).

3.3. Every $\bar{\sigma} \in \bar{\Sigma}$ is the environment of some $\sigma \in \Sigma$.

A 4: State space

4.1. For every ordered pair $\langle \sigma, \bar{\sigma} \rangle \in \Sigma \times \bar{\Sigma}$ there exists an $\mathfrak{H}$ such that $\mathfrak{H}$ is a Hilbert space.

4.2. The points ψ of $\mathfrak{H}$ are complex valued functions on $\Sigma \times \bar{\Sigma} \times E^3 \times T$.

4.3. For every $\langle \sigma, \bar{\sigma} \rangle \in \Sigma \times \bar{\Sigma}$, $\partial \psi / \partial t$ and $V^2 \psi$ are everywhere finite and piece-wise continuous.

4.4. For all $\sigma \in \Sigma$ and all $\bar{\sigma} \in \bar{\Sigma}$, if the spatial region accessible to the compound system $\sigma \dotplus \bar{\sigma}$ is $V \subseteq E^3$, then $\psi(\sigma, \bar{\sigma}) = 0$ on the border ∂V.

4.5. For every $\psi \in \mathfrak{H}$ and at any $t \in T$, ψ and $V\psi$ are square integrable over every region accessible to the compound system $\sigma \dotplus \bar{\sigma}$.

A 5: Dynamical variables

5.1. $\{Q\}$ is a nonempty family of functions on Σ.

5.2. $\{\mathcal{Q}\}$ is a ring of operators in $\mathfrak{H}$.

5.3. If $Q \in \{Q\}$ is a property of $\sigma \in \Sigma$, then there exists a $\mathcal{Q} \in \{\mathcal{Q}\}$ such that $\mathcal{Q}$ represents Q.

5.4. For every $\sigma \in \Sigma$, every $x \in E^3$, every $t \in T$, every $Q \in \{Q\}$ and every $\mathcal{Q} \in \{\mathcal{Q}\}$,

(a) $\mathcal{Q}$ is a linear mapping of $\mathfrak{H}$ into itself.

(b) $\mathcal{Q}$ has a complete set $\{\varphi_k\}$ of orthonormal eigenfunctions φ_k defined on $\Sigma \times E^3$:

$$\mathcal{Q}\,\varphi_k = q_k\,\varphi_k, \qquad (\varphi_k, \varphi_{k'}) = \delta_{kk'} \tag{1}$$

$$\psi = \sum_k c_k\,\varphi_k, \qquad c_k : T \to C, \tag{2}$$

where 'C' designates the complex field.

(c) For every fixed $\sigma \in \Sigma$, every $\varphi_k \in \{\varphi_k\}$ is piece-wise continuous w.r.t. x.

(d) For every $\sigma \in \Sigma$, every $Q \in \{Q\}$ and every $\mathcal{Q} \in \{\mathcal{Q}\}$, if $\mathcal{Q}$ represents Q, then: eiv $\mathcal{Q} \equiv q_k \in R$.

5.5. The value of $\psi^* \mathcal{Q} \psi$ at $\langle \sigma, \bar{\sigma}, x, t \rangle$ represents the Q-density of σ when associated with $\bar{\sigma}$.

5.6. For every $\sigma \in \Sigma$ and every $\mathcal{Q} \in \{\mathcal{Q}\}$, the eigenvalues q_k of $\mathcal{Q}$ are the sole values of the property Q that σ takes on, provided $\mathcal{Q}$ represents Q.

A 6: Kind and structure of system

6.1. For every pair $\langle \sigma, \bar{\sigma} \rangle \in \Sigma \times \Sigma$ there exists an $H \in \{\mathcal{Q}\}$ which is a function of x, t and V.

6.2. Every H is invariant under spatial rotations, parity transformations, and time inversions.

6.3. For every pair $\langle \sigma, \bar{\sigma} \rangle \in \Sigma \times \bar{\Sigma}$, the eigenvalues of $H(\sigma, \bar{\sigma})$ represent the energy values of σ when acted on by $\bar{\sigma}$.

A 7: Evolution of states

7.1. If $H \in \{\mathcal{Q}\}$ and $\psi \in \mathfrak{H}$ refer to one and the same pair $\langle \sigma, \bar{\sigma} \rangle \in \Sigma \times \bar{\Sigma}$ then: for every $t \in T$ and every $x \in E^3$,

$$i \hbar \frac{\partial \psi}{\partial t} = H \psi. \tag{3}$$

7.2. If, for a given $\langle \sigma, \bar{\sigma} \rangle \in \Sigma \times \bar{\Sigma}$, there exists a $\psi \in \mathfrak{H}$ that solves equation (3), then: up to an arbitrary phase $e^{i\alpha}$ with $\alpha \in R$, $\psi(\sigma, \bar{\sigma})$ represents the state of σ at time t when acted on by $\bar{\sigma}$.

7.3. If $U \in \{\mathcal{Q}\}$ is a unitary operator in $\mathfrak{H}$ and $\psi \in \mathfrak{H}$ and $\langle \sigma, \bar{\sigma} \rangle \in \Sigma \times \bar{\Sigma}$, then $\psi(\sigma, \bar{\sigma})$ and $U \psi(\sigma, \bar{\sigma})$ represent the same state of σ when associated with $\bar{\sigma}$.

7.4. $\hbar$ is a positive real number such that $[\hbar] = L^2 M T^{-1}$.

This completes our list of comprehensive axioms. They specify only a part of the primitives and even so incompletely — thus, so far the available information is insufficient to effectively solve the Schrödinger equation (3). The characterization of the primitives will be completed by the specific axioms in Sec. 7.

5. Some comments

The mathematical postulates are well-known and they can be refined at various points. Our main concern is with the nonmathematical ones, i. e. with those bearing on the physical meaning of the basic symbols [6]. We will not dwell on E^3 and T because protophysics — the set of generic physical theories presupposed by most specific theories — takes care of them [1].

The central concept of QM is the one of quanton, here symbolized Σ; it is central because QM focuses on Σ. Thus every dynamical variable represents some objective (but possibly unobservable) property of a $\sigma \in \Sigma$. But in most cases σ occurs in the company of $\bar{\sigma}$, the classically describable environment of σ. Eventually $\bar{\sigma}$ may be nonexistent (the null individual of the kind Σ). In this case σ will be a free quanton. (The mere occurrence of this concept shows that QM is not primarily about quanton-measuring device complexes.) But in most cases the quantum-mechanical formulas refer to (are about) quanton-environment pairs. In other words, the reference class of QM is the cartesian product $\Sigma \times \Sigma$. In short, QM is about physical systems not about appearances, measurement acts, observers' decisions, theorists' uncertainties, and so forth. Moreover, axiom $A\,3.1$ exhibits our commitment to the philosophical hypothesis that there is an external world constituted by σ's and $\bar{\sigma}$'s.

The next semantical component or correspondence "rule" is $A\,5.5$, which enjoins us to take the densities $\psi^*\,\mathcal{Q}\,\psi$ rather than the corresponding operators for the primary physical magnitudes of QM. Thus we shall not speak of the position but of the position density (or distribution in x) of a quanton, if only because in the coordinate representation x is not a particle coordinate defined on $\Sigma \times K \times T$, K being the set of reference frames.

Another meaning-giving postulate is $A\,5.6$, which tells us what values a property can really take on. The usual interpretation is that the eigenvalues q_k of $\mathcal{Q}$ are the sole possible results of an ideal measurement of Q on σ. We have dropped this interpretation not only because physical theories are supposed to say what the world looks like even when nobody is looking, but also because of the following technical reasons. Firstly, a quanton rather than a quanton-*cum*-apparatus is here at stake as shown by $A\,5.1$. Secondly, ideal measurements have only a heuristic and didactic use. Thirdly, only exceptionally does a run of real measurements yield an exact eigenvalue. Fourthly, we should not prejudge the issue: prophets, not scientists, are called to say what the sole possible outcomes of a human operation can be.

The last distinctly semantical components of our axiom system are $A\,7.2.$ and $A\,7.3.$ The former says that every state of a quanton in an environment is representable by a ray in Hilbert space, that satisfies the Schrödinger equation. The other postulate says that there are infinitely many mathematical representatives of any given physical state: it is a postulate of objectivity, one that presupposes the existence of a unique real state referred to by infinitely many different "wave" functions. In this regard $A\,7.3$ is akin to the covariance principles, all of which suggest a unique concrete and autonomous thing behind an infinite set of conceptual representations.

6. Some comprehensive theorems

A first theorem, or rather metatheorem, that can be deduced from the above axioms is this. If the pair $\langle \sigma, \bar{\sigma} \rangle$ is represented by ψ, and the property Q of σ is represented by $\mathcal{Q}$, then the probability that Q lies in the eigenvalue interval $[q_{k'}, q_{k''}]$ is

$$w\big(q_k \in [q_{k'}, q_{k''}]\big) = \sum_{k \in \Delta k} |c_k|^2, \qquad \Delta k = [k', k''] \tag{4}$$

where c_k is the kth component of ψ along the axis $\varphi_k = \text{eif } \mathcal{Q}$ in the Hilbert space of $\langle \sigma, \bar{\sigma} \rangle$. The proof consists in showing that the w's satisfy KOLMOGOROFF's axioms for probability — which they clearly do.

This result is important because it is usually assumed as an extra postulate, which leaves the uneasy feeling that there could be alternative, nonstochastic interpretations of the usual formalism of QM [7]. We can now assert that BORN's probability interpretation is not *ad hoc* — provided a physical not a psychological interpretation of probability is adopted. Surely, if new primitives are added which fail to scatter, then a half-stochastic theory can be obtained in which probability distributions are derived; but since the primitive base is changed, the whole theory is different from the usual QM rather than being just a reinterpretation of the usual formalism.

Another generic theorem is this. When the pair $\langle \sigma, \bar{\sigma} \rangle$ is in the state represented by ψ, the average (not the expectation value) of the property Q represented by $\mathcal{Q}$ equals

$$\langle \mathcal{Q} \rangle_\psi = \sum_k |c_k|^2 \, q_k = (\psi, \mathcal{Q}\,\psi). \tag{5}$$

As a corollary we obtain the usual expression for the standard deviation of the Q distribution. Eq. (5) and its corollary complete the statistics of an arbitrary property. So far nothing need or can be said about the statistics of a (finite) set of measurement results: our averages and scatters are theoretical, objective and moreover measurement-free. Of course *if* QM is true then we should find *nearly* the same results upon performing the appropriate measurements. But this requires several additional hypotheses — e. g., that the measured frequencies will fluctuate about the calculated probabilities. In any case, the interpretation of the root mean square $\Delta_\psi \mathcal{Q}$ as a subjective uncertainty is nonphysical. Such an interpretation is a leftover from the 18th century interpretation of probability as a measure of ignorance, as well as from the classical picture of quantons as point particles. We are not justified in feeling uncertain about the sharp position of something that normally fails to have a definite location, although it may be manipulated (prepared) to obtain one.

Once we have the above theorems we deduce the general Heisenberg relation for two arbitrary properties represented by operators $\mathcal{Q}_1$ and $\mathcal{Q}_2$ subjected to the condition $\mathcal{Q}_1 \mathcal{Q}_2 - \mathcal{Q}_2 \mathcal{Q}_1 = i \mathcal{Q}_3$:

$$\Delta_\psi \mathcal{Q}_1 \cdot \Delta_\psi \mathcal{Q}_2 \geqq \tfrac{1}{2} |\langle \mathcal{Q}_3 \rangle_\psi| .$$

Notice that no apparatus is so far in view. We do not even know what properties the operators $\mathcal{Q}_1$ and $\mathcal{Q}_2$ stand for, and it is impossible to build general purpose measuring devices.

7. Specific Postulates

We shall now specify the form and content of the remaining primitives by stipulating, among other things, that the quantons of this theory are endowed with mass and eventually also charge.

A 8: Position and momentum densities

If $\sigma \in \Sigma$, when associated with $\bar{\sigma} \in \Sigma$, is in the state represented by ψ, and if $x \in E^3$, then:

8.1. The instantaneous position density of σ is $\psi^* \, x \, \psi$.

8.2. The instantaneous momentum density of σ is $\psi^* (\hbar/i) \, \nabla\psi$.

A 9: Mass and charge

9.1. $\mathcal{M}$ and $\mathcal{E}$ are both functions from Σ into the non-negative reals.

9.2. The value μ of $\mathcal{M}$ at σ represents the mass (inertia) of σ.

9.3. The value e of $\mathcal{E}$ at σ represents the electric charge of σ.

A 10: Nonelectromagnetic forces

10.1. V is a function of $\sigma, \bar{\sigma}, x$, and ∇.

10.2. For any given pair $\langle \sigma, \bar{\sigma} \rangle \in \Sigma \times \Sigma$, $\psi^* V \psi$ at $x \in E^3$ represents the instantaneous density of the nonelectromagnetic actions of $\bar{\sigma}$ upon σ.

A 11: Electromagnetic potentials

11.1. $\langle A_0, A \rangle$ is a real valued 4-vector field on $\Sigma \times E^3 \times T$.

11.2. $\langle A_0, A \rangle$ satisfies MAXWELL's second order equations and the gauge condition.

11.3. $E = -\nabla A_0 - \partial A/\partial(ct)$ and $B = \nabla \times A$ represent the electric and magnetic field strengths respectively.

A 12: Hamiltonian of quanton in external field.

If $\sigma \in \Sigma$ and $\mathcal{M}(\sigma) = \mu \neq 0$ and $\mathcal{E}(\sigma) = e$ and if the action of $\bar{\sigma}$ on σ is represented by V, A_0 and A, then:

$$H = (1/2\,\mu) \left(\frac{\hbar}{i} \nabla - \frac{e}{c} A \right)^2 + \frac{e}{c} A_0 + V . \tag{6}$$

This completes our axiomatic formulation of QM. It is surely perfectible but at least it supplies a rough characterization, both mathematical and physical, of the basic concepts and it entails all the usual theorems. Some such consequences will be recalled in the next section. Let us now make a few comments on the preceding axioms.

We have abstained from stating that x represents the position and $\frac{\hbar}{i} V$ the momentum of σ, and this because "position" and "momentum" are classical concepts which have no place in QM except as classical limits. In this theory x is just a name for an arbitrary point in configuration space, not a particle coordinate, while $(\hbar/i) V$ is in itself physically meaningless. Only their eigenvalues, densities and averages have a physical meaning. And to compute them we need the previous postulates and the mathematics lying in the background, which mathematics has classical logic built into it. Consequently the whole point of the so-called quantum logic vanishes in thin air. Indeed, in QM we do not even pose the questions "At what point is σ?" and "With which speed is σ moving when passing through the point x?", for the concepts of sharp position and momentum of a particle are neither among the primitive nor among the defined concepts of the theory. And they do not occur in it because we are not assuming that quantons are either randomly moving point particles or orderly propagating fields: quantons are utterly *sui generis* entities and this is why a special theory is needed to account for them.

8. Some Special Theorems

A first consequence of the foregoing is Born's original postulate, which we restate as follows. For any pair $\langle \sigma, \bar{\sigma} \rangle \in \Sigma \times \Sigma$ in a state $\psi \in \mathfrak{H}$, the probability that σ lies in a region $V \subseteq E^3$ is

$$\int\limits_{V} d^3 x\, |\psi(x)|^2.$$

In the usual interpretations this is the probability that the *position* of σ will be *found* to lie within V upon performing a position *measurement* on σ. But this interpretation is *ad hoc* and moreover blatantly false. First, nothing has so far been said about measurement devices. Second, the probability of finding something somewhere is conceptually and in general also numerically different from the probability of that something being there, whether this probability is interpreted as a relative frequency or as a measure of the propensity for the thing to dwell in the given region. Thus without the adequate techniques we will fail to find anything we care to mention, be it a galaxy or a certain molecule in the tip of the reader's nose. Finally the usual interpretation of Born's theorem is couched in a language which is not strictly physical. In short de Broglie's phrase *probabilité de présence* is hereby vindicated.

Another logical consequence is of course HEISENBERG's scatter relation for x and $(\hbar/i)\ \nabla$. In the present interpretation they are, as suggested long ago [8], objective scatter relations. Only, they refer now to individual quantons, whether free or under the action of an environment, rather than to statistical aggregates of either similar quantons or similar observations on a single quanton. Needless to say, *if* QM is true then the scatters computed for the individual quanton must be close to the values obtained upon making measurements on such aggregates. Unfortunately no exact computations for real experimental arrangements, e. g. the single finite slit, let alone for the double slit, have ever been performed. One believes the Heisenberg relations because they are entailed by postulates which entail lots of experimentally confirmed theorems, not because they nicely fit imaginary experiments described qualitatively in the first chapters of textbooks but occurring nowhere else.

In addition to these and several other theorems, certain metatheorems are proved. One of them is the assertion that SCHRÖDINGER's equation is invariant under a gauge transformation of the first kind (phase shift) provided the electromagnetic potentials are subjected to a gauge transformation of the second kind. Let us hope no operationalist will waste his time trying to confirm this metatheorem in the laboratory. Yet anything is possible under the action of an antitheoretical drug.

In any case, if a quantum theory of measurement is wanted, then certain assumptions and definitions must be added. Such a theory is sketched elsewhere. In accordance with MARGENAU's recommendation [9], our theory dispenses with VON NEUMANN's projection postulate, which is just the command "Laws of nature and in particular SCHRÖDINGER's equation: stand still for here I come, I the omnipotent Observer, I the supreme magician who can conjure up anything I wish to". Yet our theory of measurement has the same fatal shortcoming of the usual one: it is generic and therefore it concern no real measurement. Every real measurement involves a number of specific physical laws, some of which are classical, whence all realistic theories of measurement must be specific and must involve a number of physical theories. In any case, for a typical static interaction we obtain the soothing result that two successive measurements of a given property have the same probability — but, thank heavens, not the same value.

9. Concluding Remarks

In order to dispel the fog that surrounds QM one need neither return to classical mechanics nor resort to nonclassical logics. (Moreover one must not do the latter (a) because the mathematics used by QM has classical logic built into it and (b) because if QM were reformulated on

8*

the basis of some system of nonclassical logic, which has not been done, then it would isolate itself from the rest of physics and would consequently become untestable, for every experimental test calls for classical theories.)

To chase away the ghosts all one has to do is (a) to uncover the formalism of QM and (b) give it a thoroughly physical interpretation. This interpretation is attached to the formalism by assigning the primitive symbols a physical meaning. And this specification of meaning is made by having those symbols stand for physical entities and properties thereof, not mental states or acts of apperception. What one gets by proceeding in this way is a formulation of QM with the following characteristics:

(1) *logical organization* (axiomatic foundation),

(2) *consistency*, both formal (freedom from contradiction) and semantical — i. e. avoidance of interpretation in terms of concepts extraneous to either the primitive base or to the bunch of terms of the physical jargon occurring in the semantic characterization of the primitives;

(3) *thoroughly quantal*: no point particles and no waves;

(4) *thoroughly physical*: no observer-dependent features.

Ours is of course just one possible physical axiomatization of QM. Alternative axiomatic foundations should be welcomed. In particular, it would be interesting to have a subjectivistic axiomatization of QM, especially since v. Neumann's is neither axiomatic nor consistent [1] although it passes for being both. Such a subject-centered axiomatization should of course start by postulating the characteristics of The Observer and The Theoretician: i. e., it should be a piece of psychology. And every physical entity, property and relation should be characterized in psychological terms, i. e. it should be defined in terms of the subject's perceptions and thoughts. This has not been done although it is frequently believed that the usual formulation of QM deals consistently with the observer and his operations. Moreover it could be argued that no such subjectivistic foundation of QM can be achieved for, after all, observers and what they observe are made up of quantons, not the other way around.

In any case, the next move is up to the subjectivist: it is up to him to challenge the realistic interpretation of QM and (a) to exhibit a consistent formulation of QM based on perceptions and apperceptions, and (b) to prove that such an interpretation is preferable to the realistic one, not only from the point of view of his philosophy but also scientifically, in the sense that it facilitates our understanding of nature. Meanwhile let us enjoy the commerce with the external world [10].

References

[1] BUNGE, MARIO: Foundations of physics. New York: Springer 1967.
[2] — Strife about complementarity. Brit. J. Phil. Sci. 6, 1, 141 (1955); — The philosophy of the space-time approach to the quantum theory. Methodos 7, 295 (1955); — A survey of the interpretations of quantum mechanics. Am. J. Phys. 24, 272 (1956); — Physics and reality. Dialectica 19, 195 (1965).
[3] NOLL, W.: Space-time structures in classical mechanics. In: M. BUNGE (ed.), Delaware seminar in the foundations of physics. Berlin-Heidelberg-New York: Springer 1967.
[4] POPPER, K. R.: The propensity interpretation of probability. Brit. J. Phil. Sci. 10, 25 (1959).
[5] HOUTAPPEL, R. M. F., H. VAN DAM, and E. P. WIGNER: The conceptual basis and use of the geometric invariance principles. Revs. Mod. Phys. 37, 595 (1965).
[6] BUNGE, M.: The structure and content of a physical theory, in the collective volume cited in Ref. 3.
[7] FEYNMAN, R. P., and A. R. HIBBS: Quantum mechanics and path integrals. New York: McGraw-Hill Book Co. 1965.
[8] POPPER, K. R.: Die Logik der Forschung. Wien: Springer 1934. Last rev. ed.: The logic of scientific discovery. New York: Harper & Row 1965.
[9] MARGENAU, H.: Quantum-mechanical description. Phys. Rev. 49, 240 (1936).
[10] For further discussions on the subjectivism-objectivism issue in QM, see references 1 and 2, and Quanta and philosophy, 7e Congrès inter-américain de philosophie. Québec: Presses de l'Université Laval 1967.

Studies in the Foundations
Methodology and Philosophy of Science

Volume 1

Delaware Seminar in the Foundations of Physics

With contributions of PETER G. BERGMANN, PAUL BERNAYS, MARIO BUNGE, HAROLD GRAD, EDWIN T. JAYNES, PETER HAVAS, HENRY MARGENAU, WALTER NOLL, JAMES L. PARK, E. J. POST, RALPH SCHILLER, CLIFFORD A. TRUESDELL

With 5 figures. XII, 193 pages 8vo. 1967. Bound DM 38,— / $ 9.50

Volume 3

Scientific Research

By MARIO BUNGE

3/I. The Search for System
With 64 figures. XII, 536 pages 8vo. 1967. Bound DM 84,— / $ 21.—

3/II. The Search for Truth
With 61 figures. VIII, 374 pages 8vo. 1967. Bound DM 68,— / $ 17.—

Springer Tracts in Natural Philosophy

Volume 10

Foundations of Physics

By MARIO BUNGE

With 5 figures. XII, 311 pages 8vo. 1967. Cloth DM 56,— / $ 14.—

MIX
Papier aus verantwortungsvollen Quellen
Paper from responsible sources
FSC® C105338

If you have any concerns about our products,
you can contact us on
ProductSafety@springernature.com

In case Publisher is established outside the EU,
the EU authorized representative is:
**Springer Nature Customer Service Center GmbH
Europaplatz 3, 69115 Heidelberg, Germany**

Printed by Libri Plureos GmbH
in Hamburg, Germany